The Story
of Antimatter
Matter's Vanished Twin

The Story of Antimatter

Matter's Vanished Twin

Guennadi Borissov

Lancaster University, UK

World Scientific

EW JERSEY · LONDON · SINGAPORE · BEIJING · SHANGHAI · HONG KONG · TAIPEI · CHENNAI · TOKYO

Published by

World Scientific Publishing Co. Pte. Ltd.

5 Toh Tuck Link, Singapore 596224

USA office: 27 Warren Street, Suite 401-402, Hackensack, NJ 07601

UK office: 57 Shelton Street, Covent Garden, London WC2H 9HE

Library of Congress Cataloging-in-Publication Data

Names: Borissov, Guennadi, author.

Title: The story of antimatter : matter's vanished twin / Guennadi Borissov
(Lancaster University, UK).

Description: Singapore ; Hackensack, NJ : World Scientific Publishing Co. Pte. Ltd., [2018] |
Includes index.

Identifiers: LCCN 2017046015| ISBN 9789813228757 (hardcover ; alk. paper) |
ISBN 981322875X (hardcover ; alk. paper)

Subjects: LCSH: Antimatter. | Particles (Nuclear physics)

Classification: LCC QC173.3 .B67 2018 | DDC 530--dc23

LC record available at https://lccn.loc.gov/2017046015

British Library Cataloguing-in-Publication Data

A catalogue record for this book is available from the British Library.

For any available supplementary material, please visit
http://www.worldscientific.com/worldscibooks/10.1142/10673#t=suppl

Printed in Singapore

To all whom I love and who loves me

Acknowledgments

To prepare this book, I extensively used the following reviews and scientific publications: Gary Steigman, "*Observational tests of antimatter cosmologies*", Ann. Rev. Astron. & Astrophys. 14, 339 (1976); James M. Cline, "*Baryogenesis*", arXiv: hep-ph/0609145; K.S. Babu *et al.*, "*Baryon Number Violation*", arXiv: 1311.5285 [hep-ph]; Pasquale Di Bari, "*An introduction to leptogenesis and neutrino properties*", Contemp. Physics, 53 (2012) and arXiv: 1206.3168 [hep-ph], and I would like to express my gratitude to all these authors.

I also utilized the following internet resources: wikipedia.org, pdg.lbl.gov, inspirehep.net, and arxiv.org, and I thank all people, who develop and support these web sites. My special thanks are to the search engine Google, which considerably facilitated the search of required information.

I want to thank Lerh Feng Low, who initiated the work on this book and provided many useful comments during its preparation, and Rebecca Yap, whose comments and questions helped me to improve the content of the book. I would also like to thank Anna Borissova for her significant contribution in the improvement of the text, and Alla Dubrovina, who prepared many illustrations for this book. My sincere thanks are due to Natalia Ozhogina, who read the initial version of the book and provided many useful comments and suggestions. I am indebted to Rachel Seah Mei Hui for the excellent editorial support of this book.

Contents

PART 1

Particles and Antiparticles

"Any scientist who can't explain to an eight-year old what he is doing is a charlatan."
Kurt Vonnegut, *Cat's Cradle*

Chapter 1

The Matter of Antimatter

In this world, everything has its causes and its effects. Generations come, conceive new generations and fade into oblivion. Tectonic plates collide with each other, rear up and create mountains. The force of gravitation gathers stardust, condenses it into nebulae and forms planets. But the cause of one colossal and inconceivable phenomenon is not known at all. And yet, all other events, the origin of stars, mountains, and humans, stem from its effects. It is, therefore, the primary cause for everything.

This sole phenomenon is the birth of the *universe*. Here, the word "birth" means that our wonderful world, containing a myriad of stars embedded in the vast reaches of space initially did not exist. Suddenly, an embryo of the universe appeared. Later, during the following billions of years, this embryo developed into the world, which now extends in front of us.

By itself, the concept of the universe's birth does not sound too strange. From time immemorial, people of different tribes and nations nurtured the idea of the world originating from total nothingness. The history of the universe's creation has been described in the form of legends and myths, which vividly reflect an unlimited richness of human imagination unspoiled by scientific knowledge. Some of these legends were smoothly transformed into religious beliefs and remained a part of human ideology for many centuries. In most of the religions, God the Creator brought the world into being. Hence,

3

by accepting this dogma, we must also agree that the universe initially did not exist and had materialized all 'at once. That is why conviction in the commencement of the universe is deeply rooted in the consciousness of people.

Modern science entirely agrees with ancient legends and religious philosophy in their stance that the universe did not always exist. The reason for this unexpected consonance are the facts that undeniably prove that the universe had a beginning. Until these proofs were found and understood, opposing opinions warred, and the scientific community pondered various possibilities. But almost all thus-considered hypotheses have not survived under the weight of collected evidence. Finally, only one assumption, that of the universe's birth, has remained.

According to the dominant scientific theory, the universe was created as a result of a giant burst of energy called the *Big Bang*. The origin of this burst is not yet known. Different theoretical models attempt to explain it, but experimental confirmation is still insufficient or missing. Also, the description of what happened in the first few instants after the Big Bang is mainly speculative. This uncertainty suits religion, lending support to its confidence that the creation of the universe was an outcome of divine intent.

What is known well is that, initially, the currently observable universe was smaller than an atom. Matter did not yet exist at that stage. The universe contained just some *primary radiation*, the nature of which is still unknown. Immediately after its birth, the universe rapidly increased in size. During this expansion, the primary radiation generated *fundamental particles*, the infinitely small objects that constitute the basis of all known substances. That part of the world's history is better understood because the processes, which occurred at that time, can now be reproduced and investigated in scientific laboratories. This research is the subject of a distinct branch of science called *particle physics*. Thus, the study of the smallest objects existing in nature is tied tightly to *cosmology*, the science looking into the origin and evolution of our immense universe.

Sometime after their creation by primary radiation, fundamental particles were combined into atoms. The atoms were pulled together by gravitational forces and formed stars. The stars in turn were collected into galaxies. Some parts of matter, not included in the formation of stars, were transformed into planets and, finally, at least one of these planets became a cradle of life. Such is the brief description of the evolution of the universe from its origin to the present day. This general scheme is reasonably clear and understandable, but it hides one dark spot, one big mystery, which consists of the following.

Soon after the discovery of the first particles, scientists found that nearly each of them has a twin called an *antiparticle*. Indeed, almost all properties of the particle and its antiparticle are the same. Still, the relationship between them is quite special, because they destroy each other when they collide. This process is called *annihilation*. Annihilation can produce other particles and antiparticles. But if it occurs inside a dense-enough material, all that remains in the end is a flash of light and, possibly, a few of the lightest particles called *neutrinos*, which cannot be seen and are very difficult to detect.

Annihilation always destroys matter. This phrase may sound odd and more appropriate for a science-fiction novel. It completely contradicts the classical view of matter as something that exists forever and never disappears. When a heap of logs burns away in the fireplace, it leaves nothing but a small lump of ashes. But this does not mean that the matter disappears — the carbon contained in the logs transforms into carbon dioxide gas, which is ejected into the atmosphere through the chimney, and the total amount of matter before and after the combustion never changes. That is, the initial mass of the logs plus that of the oxygen, and the final mass of the carbon dioxide gas plus the ashes are exactly equal. Annihilation means something altogether different — matter stops existing.

Because of its seeming oddness, the process of annihilation, more than any other property of antiparticles, excites the imagination and attracts the attention of many writers of science fiction. For example, in the book *"Angels and Demons"* by Dan Brown, the villains steal a small amount of *antimatter* (the substance containing antiparticles) from a scientific laboratory and threaten to annihilate it and destroy a big city. Disregarding some scientific inaccuracies in this book, it is absolutely right at one point. The annihilation of one gram of antimatter would indeed release a lot of energy, enough to devastate a big city. This power is equivalent to the explosion of the atomic bombs dropped on Hiroshima and Nagasaki. Such is the main consequence of the destruction of matter. Fortunately, the collection of one gram of antimatter would require millions of years of work in any scientific laboratory. Therefore, the terrorist threat related to antimatter should not cause fear for the next million years, at least.

In all other respects, antiparticles are similar to ordinary particles. An *antiatom* can be made of antiparticles and can exist forever like many ordinary atoms. Antiatoms can produce antimaterials, which have the same properties as the corresponding materials composed of atoms. Antistars containing antimatter would be expected to shine like regular stars and would emit light as normal stars do. Nevertheless, antimatter is absent in our world. Antiatoms can be created in laboratory conditions, but even in this case the maximum duration of their existence does not exceed a few thousand seconds. Why is it this so? The answer is rather simple. All materials around us, including our body, contain only ordinary particles, which would collide with the antiparticles of antiatoms and destroy them. Still, this quick answer prompts the next question. Why is it that all materials comprise only ordinary matter? Contrary to the previous question, this one is much more challenging. Moreover, the full answer on it is not yet known. To get the real sense of the complexity of this question, it is worth returning in time to that distant point in the history of the universe when the particles were just created.

Particles and antiparticles always appear and cease to exist in pairs. For example, the production of two particles without the corresponding antiparticles was never observed. Similarly, an antiparticle can only be destroyed by a particle. It cannot vanish by itself. This *rule of pairs* does not follow from any law of nature. However, all known processes that involve fundamental particles strictly obey it. There are some theoretical models that allow a violation of this rule. However, none of these models has yet been confirmed by an experiment. On the contrary, numerous thorough experimental studies perfectly agree with the rule of pair production and destruction of particles and antiparticles.

The rule of pairs, if it is indeed valid, has a simple and inevitable consequence. Initially, matter and antimatter in the universe must have appeared in equal amounts. For each particle of matter surrounding us, there should be its corresponding antiparticle. The amount of matter present in the universe is huge. To agree with this claim, we do not need to go through the tedious process of counting all the billions of galaxies, which can be observed in the universe, multiplying this number by the number of stars in each galaxy and after that, multiplying the result by the mass of the average star. It is sufficient just to go somewhere far from city lights on a clear and moonless night; for example, in the mountains or on a boat in the sea; to raise up one's head and gaze out at the space full of splendid stars. Indeed, space is endless, and the matter that fills it is countless. A simple piece of logic supported by the knowledge of particles' properties inevitably tells us that exactly half of this magnificence should be made of antimatter. Somewhere in the vast cosmos, there should be antiworlds containing billions of antistars. But this is not quite true. Antiworlds do not exist. The amount of antimatter in the universe is not just smaller than what would be expected from the rule of pairs. Antimatter is totally absent. This statement is not a hypothesis. We will present later its experimental proof.

Hence, there is a clear contradiction in our understanding of nature. On the one hand, particles and antiparticles were produced in pairs at the earliest stage of the universe's formation. This process should result in equal proportions of matter and antimatter in the modern world. On the other hand, there is no antimatter anywhere. This contradiction needs to be resolved, and one possibility to do this is to find the cause of the disappearance of antimatter.

The quantity of missing antimatter is huge. It is not a question of one antiparticle or even a billion of them. The amount of antimatter equivalent to all matter collected in the entirety of the stars and galaxies in the whole enormous universe is missing. Why did it happen? It is still one of the deepest secrets hidden from us. The only way to get to the bottom of this mystery is to study particles. Their properties contain something special, something that allows for the complete destruction of antimatter without its twin being affected. We need to find and understand this "something".

Of course, before attempting to crack the mystery of vanished antimatter, we need to convince ourselves that this is really important. Indeed, why does antimatter matter? Why should we care about its loss?

We could simply answer that mere curiosity has always been an inherent feature of humankind. We could also state that the absence of antimatter is not just one of a multitude of scientific facts pleasing only a narrow circle of experts. This disappearance is one of the most crucial factors which shaped the structure of the entire universe and ultimately determined our existence. We will discuss our relationship with antimatter later. But if such motivations are not sufficient, there are more practical reasons for studying antimatter. One of such arguments is akin to the famous explanation given by the English physicist Michael Faraday, to justify the worth of his research. He gave it some one hundred and fifty years ago, but his reasoning is still pertinent.

Faraday made several significant discoveries related to electricity and magnetism. Among many other inventions, he built the first electric engine. He often made public demonstrations of his experiments with electricity. Purportedly, William Gladstone, then British Chancellor of the Exchequer, had been present at one of these demonstrations and had asked Faraday: "But after all, what use is it?" Faraday found an excellent answer: "Why, sir, there is every probability that you will soon be able to tax it!" This historical anecdote not only shows that physicists worry about taxes like all other people. It has a more profound meaning.

Fundamental research aims to understand the internal arrangement of nature and is not directly related to the present-day needs of civilization. So, it may sometimes look like a futile waste of time and money, with no tangible outcome. The benefits of fundamental research are usually revealed only after many years. But in any case, new understanding, gained by such study, more often than not leads to revolutionary leaps in the advance of humankind. This surge happened with electricity and magnetism. Electricity may have looked like a purposeless source of amusement to a few scientists just one hundred fifty years ago. Today, it has become a solid foundation of human civilization. It is likely that the study of antimatter will also produce such results, sooner or later. Its outcome is impossible to predict, but the knowledge obtained will certainly bring the civilization of humanity to new summits in its exploration of nature.

The investigation of antimatter has already brought about numerous achievements. Many incredible phenomena, the existence of which could not even be imagined, have been discovered. Of course, much more remains to be found. Scientists are still far from explaining why antimatter is absent. But the advances in knowledge to answer this question are substantial and deserve special consideration. This is exactly the subject of this book. It is about antimatter, its role in the evolution of the universe and its mysterious disappearance. This book aims to paint a broad picture of the active and relentless research performed in this field. It follows the leading cohort of physicists in their quest to investigate missing antimatter, branches out to the frontiers of their activity and presents the significant progress already made in understanding this fascinating phenomenon.

Reading the book does not require any special knowledge or skills. What is needed is just common sense, a chunk of imagination and a desire to learn something new. Equations, which appeal so much to scientists, but which are mostly incomprehensible to other people, are excluded from this book. All scientific ideas and results of research, even the most difficult ones, are explained in simple terms using well-known examples from everyday life. Of course, this story, like any other discussion of scientific subjects, contains some special words and expressions, like "neutrino" or "antiparticle." The inclusion of these terms is unavoidable because science always deals with unusual phenomena, which need new words to describe them. In this book, all such terms are defined and explained. Besides, they are collected in the index at the end of the book, so that their definition can be easily found. Otherwise, the content of this book should be understandable to everybody interested in modern science and its latest achievements.

Therefore, treat yourself to a cup of tea or coffee, sit comfortably in a chair, or on a sofa, or even on a train, and start reading. Our story of enigmatic antimatter begins.

Chapter 2

Big Expansion of the Universe

2.1 Receding Galaxies

The universe was born about 14 billion years ago. This unique event, which was never repeated in nature or reproduced in a scientific laboratory, is called the *Big Bang*. However, the selected term is very misleading. A bang is usually associated with an explosion, and explosions do not create — they destroy. In fact, the term "Big Bang" was invented by the English astronomer Fred Hoyle, who strongly opposed the corresponding theory describing the creation and evolution of the universe. This opposition can explain the inconsistency between the name and the content of the theory, given that Hoyle could hardly be accused of any intention to make the theory of the Big Bang clearer.

Each of us has seen an explosion, at least in a movie, and therefore can easily picture the birth of the universe in this way. It can be imagined as a giant firework exploding somewhere in the depths of the cosmos (Fig. 2.1). After such an event, stars would begin to move in all directions from the center of the explosion, like the glittering firework sparks in the darkness of night. This picture may look very impressive and worthy of being painted by artists, but it is too far from the truth. To understand why this is so, we should consider the Big Bang theory in more detail.

For a very long time, all scientific deliberations about the evolution of the universe were speculative rather than based on concrete

Fig. 2.1   The firework as an incorrect model of the Big Bang. Image courtesy of http://www.photos-public-domain.com

information. Stars shining in the night sky seemed to remain the same from year to year and from century to century. There was no other source of knowledge about possible changes in the cosmos. Only sometimes, rare events, like the passage of a comet or the explosion of a supernova, hinted to the ancient astronomers that the world beyond the earth's atmosphere was not as static as it seemed to be.

The situation rapidly changed at the beginning of the twentieth century, when very powerful telescopes were built, and methods for astronomical measurements became quite sophisticated. At that time, scientists started to notice changes occurring in outer space and to derive from them conclusions on the development of the universe.

One of such telescopes was located at the Lowell Observatory in Arizona, USA, and the astronomer Vesto Slipher used it in 1912 to measure the velocity of the so-called spiral nebulae. At that time it was not yet known that these nebulae are in fact massive aggregations of billions of stars called *galaxies*. Nobody could imagine the immense distance between them and Earth, which even light needs millions of years to cover. These galaxies were initially observed as small spots in the starry sky because the individual stars within them could not yet be discerned. Therefore, they were believed to be something like a condensation of interstellar gas and were named "nebula" (meaning in Latin "cloud") to match this belief.

Nonetheless, some of the galaxies did have a distinctive feature, which could be seen even with the old telescopes available at the beginning of the last century. They had a spiral shape reminiscent of the funnel on the surface of water draining from a bath. This particular form helped Slipher to select the nebulae for his study. Gradually collecting data and rectifying his measurements, Slipher found that most of the spiral nebulae receded from Earth at very high velocities. This result seemed to fly in the face of common sense.

First, the nebulae moved much faster than the stars surrounding them, and their elevated speed was entirely unexpected. For example, the speed of one nebula was found to be about 1100 km per second. This is an enormous value even on cosmic scales. It is much greater than the speed of the Earth around the Sun and that of the Sun rotating around the center of our galaxy, the *Milky Way*.[1] No macroscopic object can move so fast on the surface of our planet or even near it. Therefore, observing a nebula traveling at this colossal velocity was as unusual as it would be to see a cloud in the sky moving with the speed of a military jet.

Even more shockingly, the main questions were not raised only by the discovery of this tremendous velocity. Slipher's most astonishing discovery was the recession of almost all nebulae, regardless of their position in the sky. To clarify the strangeness of this observation, we can give the following simple example. Imagine watching ants somewhere in a forest. Usually, they move erratically in all directions occupied by their personal affairs and do not notice you at all. But what would you think if all the ants, as if on cue, started to move away from you? Not to the right, not to the left, not even towards you, but only directly away. You would probably decide that you had somehow caused this orderly motion. For example, you may have

---

[1]The speed of the Sun's motion relative to the center of the Milky Way is 220 km per second.

frightened the ants or be emitting some unpleasant smell or noise. Or, you may have happened upon a special spot in the forest from which some unknown force acted on the ants and drove them away. In any case, such behavior from ants would seem very strange.

Similarly, the receding nebulae inevitably prompted the question: is the Sun located in a special place in the universe, at some central point from which the nebulae move away in all directions like the sparks of a firework? The Sun is, after all, quite an ordinary star. It has only one minor feature which makes it different from the billions of other stars. The third planet of the Sun is inhabited by curious humans who gaze with admirable perseverance into the starry space. Of course, for us, this feature is not minor, but surely it could not influence the motion of the celestial objects. Nevertheless, what else, if not for the fact that we are located at the center of the universe, could explain the recession of the nebulae instead of a randomness of motion in all directions? Initially, there was suspicion of an error in Slipher's measurements, but the studies of other astronomers quickly dismissed it. Soon after the release of the first results by Slipher, no doubt remained that the nebulae do indeed move away from Earth. This fact could not remain unexplained.

The explanation came very quickly. At that time, the locomotive of science was progressing very rapidly, and the results achieved always ended up generating new ideas. In 1924, the American astronomer Edwin Hubble discovered a way of measuring the distance to the spiral nebulae. More precisely, he was the first to use one important property of the stars for his measurement. The property itself was discovered by another American astronomer Henrietta Swan Leavitt.

The career path of Leavitt was typical for many female scientists at the beginning of the twentieth century. Nobody expected any scientific discovery from them, and often they were considered just an extra member of the workforce. After graduating from college, Leavitt started working at the Harvard College Observatory. Her tasks included the routine measurement of the brightness of stars on photographs made by the telescope of the observatory. Women were not allowed to operate the telescope at that time. However, because of her duties, Leavitt had full access to all of the data obtained by the telescope. Using her measurements, Leavitt found an essential regular pattern in the properties of specific stars called *Cepheid variables*. These stars, like beacons on the seashore, regularly change their brightness. Cepheids are numerous in space, with the most known example being Polaris. The time between two consecutive peaks in the brightness of the Cepheids, is called the *pulsation period*. It can vary between one and fifty days. For example, the pulsation period of Polaris is about four days.

The pattern found by Leavitt allows one to determine the distance to the Cepheids. In general, the distance to celestial bodies is one of the most important physical quantities used in astronomy and cosmology. In particular, it is essential for the development of the theory of the universe's evolution. There are several methods of measuring this distance. One of them involves comparing the *luminosity* of a star to its *apparent brightness*. The luminosity of any radiant object, like a candle or a star, is defined as the amount of light emitted. Obviously, the luminosity does not depend on the distance from which the object is observed. Conversely, the apparent brightness of the object decreases as the distance to it increases. For example, a candle near us looks much brighter than it would from far away, although it puts out the same amount of light. The relationship between the luminosity, the brightness and the distance to an object is well known, so the distance to the object can be derived from its luminosity and apparent brightness. This method is particularly useful in the case of stars with distances from us that cannot be determined by any other method.

The brightness of a star can be relatively easily measured using a photograph of it. These very measurements were part of Leavitt's job at the observatory. But how can the luminosity be determined? Leavitt found the solution to this problem in the case of the Cepheids. She discovered that the luminosity of a Cepheid could be derived from its pulsation period. It turns out that there is a direct relationship between these two quantities, and knowing one of them also means knowing the other. This result is now called *Leavitt's law*. Obtaining the pulsation period is straightforward. For this, it is necessary just to observe the star for a sufficient amount of time. Once the period is known, the luminosity of the Cepheid can be reliably computed using Leavitt's law and consequently, the distance to it can be evaluated.

Levitt's law was of little use in astronomy until Hubble learned how to find the Cepheids in the spiral nebulae. With the development of astronomical instruments, doing this became possible. By applying Leavitt's law to the newfound Cepheids, he determined the distance to them and also to the nebulae which contain them. The discoveries that followed these measurements became the cornerstone of the modern understanding of the universe. They are considered to be one of the main achievements of twentieth-century science. The role of Hubble in this success is evident, but the contribution of Leavitt should not be discounted either. In one of his letters to Robert Hooke, Isaac Newton wrote: "If I have seen further it is by standing on the shoulders of Giants." This statement is also relevant to Hubble's results, but in this case, one of the giants lending her shoulders was a delicate woman. Unfortunately, when the importance of Leavitt's law became evident, she had already died. Sometimes fate can be cruel to people, and only historical accounts give everyone their due.

Hubble's results were stunning. He found that the nebulae are located at distances of many million *light-years*[2] from Earth. This result led to a complete U-turn, if not in the minds of all of humanity, then certainly in the minds of scientists. At that time, the universe was believed to be limited by the edges of the Milky Way. The stretch of our galaxy is quite large, but it is still "just" about a hundred thousand light-years. Suddenly, Hubble's work increased the estimated size of the universe by several thousand times. It was like releasing a fish, which had spent all of its time in a bowl, into the immensity of the ocean.[3] Similarly, Hubble's discovery opened up, for the first time, the sheer vastness of space in all its grandeur.

---

[2]The measurement of the distance between celestial objects in light years is standard in astronomy. A light-year is the length of the path traversed by light in one year and is equal to about ten trillion kilometers.

[3]Of course, the fish would probably not even understand what had happened to it.

The direct consequence of Hubble's results was the revolutionary conclusion that spiral nebulae are in fact huge galaxies which look like small spots just because of the enormous distance between them and us. Following Hubble's discovery, humanity's representation of the cosmos changed once and for all. Instead of a universe uniformly filled by stars, a different one emerged. This universe consists of isolated islands of galaxies with emptiness between them, extending millions of light-years in distance.

In 1929, Hubble made one more remarkable discovery. He compared the recessional velocities of galaxies to the distances to them.[4] The result was astonishing. Hubble found that the recessional speed of a galaxy is proportional to the distance to it. More precisely, the speed of a galaxy's motion away from us increases by 21 km/s per million light-years of distance. This result is known as *Hubble's law*. Also, the quantity, 21 km/s per million light-years, is known as the *Hubble constant*. It is interesting to mention that the current value of this constant differs from the result that Hubble derived himself. Due to some inaccuracy in his calculations, he obtained a much larger value of about 150 km/s per million light-years. This is a rare example of a case where a sevenfold mistake in a measurement does not influence the validity of the conclusion.

Hubble's discovery is not just one of many amusing scientific curiosities, which frequently flash in the news. For example, in 1953, newspapers reported a discovery made by scientists at the Oak Ridge Atomic Research Center in USA. The researchers found that 98% of all atoms in a human body are replaced each year, and none of the atoms remain in the body for more than five years. Technically, it means that each of us gets a new body every five years. This result is interesting and unexpected, but, even if it is indeed true, its importance is only limited to the human organism. Hubble's discovery is on a different scale of importance altogether. It is the pivotal point to the understanding of the universe's evolution, and it is tightly related to the most fundamental properties of nature.

---

[4]The distances were obtained by Hubble while the velocities were measured by Slipher and other astronomers.

The importance of Hubble's law and its role in the development
of modern theory regarding the universe's creation and evolution is
discussed in the next section.

2.2 Expanding Space

Hubble's law, as established by astronomical observations, prompts
several tough questions. Why does the recessional speed of galaxies
depend on the distance to them? What causes its increase? Newton's first law, which nobody has revoked so far, states that any
object, including a galaxy, must move with constant velocity unless
acted upon by some external force. Conversely, the change in velocity means that some external force must be affecting the object.
So, what force changes the speed of galaxies separated by millions
of light-years? The only known force operating on cosmic scales is
gravity. However, gravity acts to attract objects with mass, not repel
them. Therefore, gravity should slow down the motion of galaxies,
not accelerate it.

Such musings puzzled the scientific community soon after the
discoveries of Hubble. On the one hand, the increase of a galaxy's
speed with the increase in distance to it had been firmly established.
On the other hand, understanding this phenomenon using known
physical laws seemed impossible. The resolution of this paradox was
given by theorists, who study nature with the use of mathematics,
ink and a lot of paper. In 1915, at about the same time Slipher
measured the speed of galaxies, Albert Einstein presented a report
to the Prussian Academy of Sciences where he put forward equations
describing gravity.

These equations have become the foundation for the *theory of
general relativity*, which is currently the primary instrument used
to understand the universe. The theory explains numerous gravitational phenomena perfectly. For example, it predicts the gravitational
bending of light when it passes near massive galaxies. This effect has
indeed been observed. The prediction by this theory of the universe's
behavior is also very peculiar. It does not allow the universe to remain
in a static, unchanging form. The theory requires that the universe
either continuously contracts under gravitational forces or expands.

Einstein immediately realized this feature of his theory and worried a lot about it. For him, the idea of a changing universe seemed impossible, and he tried to circumvent this prediction by adding an additional term to his equations. Later he called these attempts his "biggest blunder".[5] However, another theorist, Alexander Friedmann of Soviet Russia, was not afraid of such a possibility. In 1922 he published his equations derived from the theory of general relativity. They predicted an *expanding universe*. The expansion in Friedmann's equations did not mean moving galaxies. It meant something different and, at first glance, completely impossible. His equations implied an increase in the space between galaxies, and this very effect was what caused the recession of galaxies from one another.

A simple model can illustrate Friedmann's idea. Let us imagine a balloon and suppose that its surface represents the space. Let us draw points on this surface, which will correspond to galaxies. Evidently, the points don't move anywhere. Let us now start to inflate the balloon. Its expansion results in stretching of the space and an increase in distance between the points. Thus, a small observer placed at one of these points would see all other points receding. The solution of Friedmann's equations leads to a similar expansion of the universe and, therefore, can explain the observed recession of galaxies.

Friedmann's results were not noticed by contemporary scientists although he published the work in a respectable scientific journal. Einstein himself was a referee of Friedmann's paper and did not find a single mistake in it. It seemed that the idea of the expanding space was so odd that it required the work of another theorist to support it. This theorist was the Belgian priest, astronomer, and physicist George Lemaître. In 1927, independent of Friedmann, who had already died by that time, Lemaître derived similar equations and came to the same conclusion: the space of the universe could expand and effect the recession of galaxies. Lemaître knew the experimental results of Slipher and developed his theory, in part, to explain them. Besides, his theory also proved that the recessional

---

[5]Nevertheless, the additional term introduced by Einstein is currently considered to be an important part of the cosmological theory of the universe.

speed must be proportional to the distance between the galaxies. Notably, he obtained this result before Hubble, who published his work only in 1929. Still, predicting a phenomenon is not the same as experimentally finding it. That is why Hubble's law is rightfully attributed to Hubble. In any case, Friedmann and Lemaître are also considered to be among the founders of modern theory of the universe's evolution.

At the time of its publication, the work of Lemaître was not understood by his fellow theorists, who considered it pure fiction. For example, in one conversation with him, Einstein supposedly said: "your calculations are correct, but your physics is atrocious". Of course, such "blessing" by the prominent scientist did not help to promote Lemaître's results. Therefore, only after the experimental discoveries of Hubble convincingly proved the ideas of Friedmann and Lemaître, did the scientific community start to treat them seriously. Only then, everybody agreed that these ideas were not atrocious, but were beautiful and could be used as the basis for understanding the world.

Thus, according to Friedmann and Lemaître, the recession of galaxies is a direct consequence of the expansion of the space between them. This expansion also successfully explains Hubble's law. Indeed, the recessional speed in this theoretical model is just an increase in distance between galaxies per unit time. If space stretches uniformly and each centimeter expands by a certain amount each second, then the larger the distance between two points, the greater the speed of expansion. A similar effect can be seen in the example of the balloon. When the balloon is inflated, two points located far from each other on the balloon surface recede much faster than two close points. In other words, the recessional speed increases proportionally with the distance between the two points. This effect exactly corresponds to Hubble's law.

The expansion of space relieves our small planet of the honorable but onerous duty of being the center of the world. Many people would, doubtless, like the idea that this center is located somewhere near Rome or Stonehenge. Fortunately, or unfortunately, there is no need for this. It turns out that if space expands, observers at any location will see galaxies recede from them like our astronomers have observed here on Earth. This assertion can be proven with rather simple calculations, which nevertheless require some specialised mathematical knowledge. But even without mathematics, the model of the balloon can validate this conclusion. Indeed, an observer at any point of the inflating balloon should see that all other points recede and none of them approach him. So, if the observer is vain enough, he could decide that he is located at the center of the balloon universe, although it is not precisely true. At any other point, another observer will see the same picture. Also, regardless of their position, all observers will measure the recession speed of other points to be proportional to their distance. The same effect happens in our universe. Hubble's law is valid for all extraterrestrial beings.

After accepting the possibility of space expansion, let us take an imaginary trip back in time. What will we see? With the reverse flow of time, space, evidently, will contract. The galaxies will approach one another, brought together by the shrinking space. Gradually the distance between the galaxies will become so small that all of them will fuse into a single giant aggregation of the matter filling all of the diminishing space. But the contraction will not end at this point. The density of matter will continue to increase, and the universe will become very hot. Suddenly, space will collapse into an infinitely small point, burying, in it, all of the matter and energy. This will be the end of our journey. We will arrive at the beginning of the world.

When did this beginning happen? The starting time of the universe is quite easy to calculate. All galaxies were initially within a single point, and now they are located at a certain distance from one another. Therefore, if the expansion speed is known, the age of the universe can be derived. This task is not much different from determining the arrival time of a train moving with known speed from point A to point B. The expansion speed of space corresponds to the Hubble constant. Using its value, the moment when all galaxies were at a single point has been calculated to occur about fourteen billion years ago. This is how the age of the universe has been determined. Nobody knows what was before this moment, but immediately after it, space started to inflate rapidly. Sometime after this inflation, matter got enough room to form stars, which were collected into galaxies. Then galaxies were pulled apart by the expanding space and gradually became separated by huge distances. Finally, after the passage of about fourteen billion years, the universe took the familiar form which we can now admire by staring into the night sky.

Such reasoning, supported by Hubble's discovery and several other studies, resulted in the construction of the theory of the universe's evolution called "Big Bang." Indeed, the presented picture could make one think of some explosion similar to the firework mentioned earlier. However, there are at least three significant differences between an explosion and the process of the universe's creation that are implied by the Big Bang theory.

First, the universe does not have any unique center from which the galaxies recede in all directions. The expansion of the universe happens each second and at each point of space. It is revealed at the cosmic scale as the recession of galaxies.

Second, at the moment of the Big Bang, contrary to an explosion (or a 'bang'), matter was not ejected into space. What happened then and what continues now is the expansion of space. The flow of space washes matter to all parts of the universe. Matter itself does not move at all.

Finally, the recessional speed of the galaxies increases with an increase in the distance between them. This increase is the direct consequence of the expansion of space. It never happens in the case of a normal explosion.

Due to these implacable differences, a much more appropriate name for the process of the universe's creation would be the "Big Expansion." Of course, the term "Big Bang" is already deeply ingrained in the minds of people, to such an extent that even a sitcom exploits it as a title. Therefore, it will probably never be changed. Nevertheless, it is important to remember that the creation of the universe was not an explosion at all, but rather, an expansion.

After answering the questions on the receding galaxies, we face new, much more challenging ones. Why does space expand? What defines its expansion speed? Did it always expand? Will it expand forever? And how is this expansion related to the mysterious dark energy? Answers to these questions could be the subject of a separate book, but dealing with them here would lead us far beyond the issue of antimatter. That is why we will instead consider another question — the origin of matter in the universe.

2.3   Energy of the Universe

Sometimes, human language does not have enough words to describe phenomena which exist in nature. In such cases, the dry language of mathematics is used. This language expresses no emotions. It employs, without any hesitation, terms like "infinitely small point." However, explaining these terms through known and understandable images or feelings can be very difficult or even, impossible. The events that occurred immediately after the universe's creation are of this kind. In mathematical language, we can freely say: "the universe comes into existence at an infinitely small point." However, human language has no way to convey this process. Does it really mean that all planets, stars, and galaxies were crammed into a volume smaller than that of an atom?

Let's try to picture the neighbouring house being compressed into the size of a grain of sand. Even without elaborating how this compression can be done and whether the inhabitants of the house have agreed to participate in such an experiment, this process is arduous to imagine. Now, let's try to picture that the entirety of our planet, with all its seas and mountains, forests and cities, animals, fishes and people, is crammed into the size of the grain of sand. This is already something entirely unbelievable. However, according to mathematics, not only our planet, and not only the Sun with all its planets, and not only our galaxy with the billions of stars in it, but all of the existing matter in the universe had been squeezed into an "infinitely small" space. How could this compression happen? How could this phantasmagorical picture be reconciled with our consciousness? Solving this riddle is definitely much harder than explaining how a camel can go through the eye of a needle.[6] But we will do our best to solve it.

The answer to this puzzle is surprisingly simple. The newborn universe did not contain planets, stars, galaxies, or even the atoms from which these stars were made. Even the fundamental particles composing the atoms did not exist in those first moments. Instead, there was a lot of *energy*. The entire universe was a clot of energy compressed into a minuscule volume. Of course, such a short answer does not explain anything and instead prompts many other questions. Above all, what is energy?

About two thousand years ago, the prefect of the Roman Empire, Pontius Pilate, asked a three-word question "What is truth?",[7] and after that, for many centuries, generations of scholars, theologians and artists tried to answer it. The question "what is energy?" is similar: it is short but tough to answer. In physics, energy is the most essential property of any object, and this importance can be summarized in just a single phrase: energy makes the world move. Without energy, nothing can exist. Both the smallest particle and the largest galaxy have it. Energy is also essential for everyday life. All of us use it and roughly understand what it means. We pay for energy to heat our homes or drive a car. However, its formal definition is not so simple because there are many different forms of energy, which do not resemble one another. The energy of any moving object is termed to be *kinetic*. Also, there are electrical, chemical, potential, thermal, nuclear, and even *dark*[8] energies. All these types are not directly linked to motion. Many people also claim to possess energy of thought, which can heal wounds and turn on electric bulbs. However, these phenomena belong to a separate facet of human activity and are not related to the content of this book.

---

[6] An allusion to the well-known biblical statement that "it is easier for a camel to go through the eye of a needle than for someone who is rich to enter the kingdom of God" (Matthew, 19:24). A similar statement is also made in the Quran (Al-A'raf, 7:40).

[7] Pilate asked this question during the interrogation of Jesus of Nazareth (John, 18:38).

[8] Dark energy is supposed to fill the entire space of the universe and influence its expansion. It is not related at all to the plot of the "Star Wars" saga.

Kinetic energy determines the capability of an object to move against an external force. Suppose we throw a ball upward. When it leaves our hands, we pass some amount of kinetic energy to the ball. This energy allows the ball to move against the gravitational attraction of Earth and climb some height. The greater the speed of the ball, the larger its energy, and therefore, the greater the height it can rise to. The other types of energy are harder to define without mathematical formulae. However, each type can be transformed, at least in principle, into kinetic energy. In this way, a relationship between them can be established.

Energy has two important properties. **First**, it cannot appear from anything or disappear into thin air. This property is known as the *law of energy conservation*. More precisely, this law states that the *total energy* of any isolated group of objects remains the same, regardless of any changes within this group. This law does not allow any exceptions, although many people in the past have tried to prove contrary. In particular, inventors of all times and from all countries dreamed of constructing the perpetual motion machine, which could produce more energy than it consumed. Of course, the creation of such a mechanism would resolve all of the energy problems of humankind, and people would no longer spill blood for oil, one of the current primary sources of energy. Unfortunately, such a perpetual motion machine contradicts the law of energy conservation, and therefore all attempts to build it are doomed to failure. Nobody and nothing can violate this powerful law, which is considered to be one of the main unbreakable principles governing the universe.

The **second principal property** of energy consists of its ability to freely change type and to pass from one object to another. Let us consider one of such chains. The nuclear energy released by the fusion of hydrogen nuclei inside the Sun transforms into light energy, which is emitted in all directions. Some of this light reaches the surface of our planet and hits silicon panels of a solar power station. In these panels, light energy is converted into electrical energy. Electrical energy is sent to our homes by wires and heats water in an electric kettle, where it turns into thermal energy. Finally, the water boils so that we can enjoy a cup of tea.

In fact, the chain of energy transformations does not start at the Sun. There is only one primary source of energy in the universe. It is the Big Bang. The energy released by this colossal phenomenon eventually reached the kettle after many conversions and exchanges between different objects, making an intermediate stop at the Sun. During this long path, the total amount of energy in the universe always remained the same. It is just transferred from one object to another, like a baton which passes from runner to runner in a relay race. Thus, sitting near a fire at a campsite, admiring the sunset at the seashore, or simply, heating water on a gas stove, we observe no more nor less than the gleams of the Big Bang coming to us through the tunnel of time and space.

Let us consider in more detail the type of energy transferred by *radiation*. There are many kinds of radiation and one of them is well known to everybody — light. Light energy coming to us from the Sun is the source of almost all other types of energy available here on Earth. Only nuclear, geothermal, and tidal energies have a different origin. Humans started to use the energy of light when they started the first fire to warm themselves and to heat their homes.

The largest part of energy released by the burning of firewood is emitted as invisible infrared light. However, both infrared and visible light are just different forms of *electromagnetic radiation*. This radiation surrounds us everywhere. Its applications are so numerous that it is impossible to imagine modern life without it. When we speak on the phone, switch TV channels using a remote control, have an x-ray test at a hospital, or listen to music on the radio, we use electromagnetic radiation. Of course, for us, the most important form of this radiation is visible light, which we use to see and explore the world.

Radiation has also been essential for the universe's creation. As was discussed already, the starting size of the universe was petite, and matter could not exist in it yet. Instead, the universe was filled with highly energetic radiation. Quite often, artists present the Big Bang in their paintings as a flash of light. But the very first radiation was not light. The universe at that time was dark but still full of energy carried by *primary radiation*. Its nature is hard to determine because primary radiation existed only during the first few instants after the universe's birth and has not been observed experimentally since. However, it is well known that the energy of this radiation was enormous. In fact, scientists think that it is due to this huge energy that the space of the universe started to expand.

According to the law of energy conservation, the total energy of the universe in the initial moment of its existence should be equal to the amount seen now. This amount can be calculated and written down on a piece of paper as a very long number, but this would not help to picture it or compare it with something familiar. The number of currently known galaxies exceeds a hundred billion. Each galaxy contains billions of stars, and each star is a giant store of energy, at least from our point of view. It is sufficient to mention that the Sun has heated the Earth and other planets of the solar system for four and half billion years and will continue to radiate energy for another seven billion years. Even without multiplication of all these numbers, it is easy to accept that the current amount of energy in the universe is gigantic. All this energy should also have been present in the newborn universe, which was tiny. It may seem impossible to fit all the energy of the universe into such a small volume. However, it turns out that any amount of radiant energy can be compressed into a volume of any size without limitations.

To better understand this property of radiation, let us consider a simple physical instrument called burning glass. Essentially, it is a lens, which focuses rays of sunlight passing through it, onto a very small area. All of the sunlight's energy is also concentrated in the same spot. The accumulation of energy heats the exposed surface and can even ignite it. Such experiments with a burning glass, which many of us conducted as children, clearly demonstrate that radiation, together with all of its energy, can be concentrated in a very small volume without much effort. Similarly, the young universe, its little size notwithstanding, should be able to contain as much energy as required for its future growth and development into its current state.

Thus, the answer to the initial question raised at the beginning of this section becomes clearer. At least, a part of it is now known. Initially, the universe could not contain any matter, but instead, had a lot of energy in the form of radiation. This answer does not yet explain how the universe could evolve to its current state from the "infinitely small point", but nonetheless, it is a giant leap forward.

Based on this knowledge, we can also attempt to solve the old physical riddle about the camel and the eye of a needle. Its solution may look like this. How can a camel go through the eye of a needle? "It's elementary, my dear Watson".[9] We just need to convert the camel into radiation. Radiation can easily go through the eye of the smallest of needles. After that, we transform radiation back into the camel, and that's it, the problem is solved.

Of course, the main idea of this solution looks simple and clear. But the practical details of its implementation are not straightforward. How do we convert the camel into radiation? And, how does radiation transform back into a camel?

A similar question arises in the case of the universe's evolution. How did invisible radiation produce matter which can be seen and even touched? Is it a kind of magic? Later, this transformation will be explained without any involvement of magic. We will also consider reverse processes when matter is converted into radiation. But before doing this, we need to review, in more detail, matter, radiation, and the tight relation between them. The starting point of this discussion is the structure of matter, which we present in the next chapter.

---

[9]This phrase, contrary to the common belief, was never said by Sherlock Holmes in the original stories by Sir Arthur Conan Doyle.

Chapter 3

Microworld of Particles

3.1 Structure of Matter

In this chapter, we start our first immersion into the fascinating *microworld of particles*. This world can be found everywhere, in each tiny grain of sand. It is around us and even inside us. But it is unimaginably minuscule and cannot be seen by the naked eye nor touched by hand. Everything is different there. Entry into this smallest of worlds is difficult. Scientists need the largest instruments ever built to penetrate it. In this book, we shall follow researchers in their quest to unravel the secrets of this wonderland. We shall walk its winding tracks, meet its unseen inhabitants, try to understand their behavior and make many unexpected conclusions. Maybe, the most surprising thing about them is that these tiny particles, regardless of their obscurity, are the real lords of the universe. They rule the destinies of huge stars and galaxies and determine our existence.

Since the beginning of time, people have always tried to understand the inner structure of the world surrounding them and to find the pieces from which it is composed. Of course, there was no lack of different opinions on this subject. According to one of them, all matter consisted of small indivisible parts called *atoms*. This idea is attributed to the Greek philosopher Democritus, who lived about two and half thousand years ago. It is fair to say that many other people from different parts of the world also put forward similar ideas. But in any case, Democritus invented the term "atom", which in ancient

Greek means "indivisible." The following course of events proved that he was entirely right regarding atoms being the building blocks of any material, but he was rather wrong concerning their indivisibility. This mistake is not surprising. At the time of Democritus and many centuries since then, there were absolutely no tools nor methods to check his suggestion. It was a figment of the imagination. It had the same right to exist as many other similarly ingenious ideas, like the statement of Aristotle that the world consists of earth, water, air, fire and an enigmatic fifth element called ether. All such hypotheses were very handy for heated disputes between fellow philosophers during leisurely walks in the olive groves of Academe. But it was difficult, and probably impossible, to convince the opponent in such a debate. Philosophers rarely changed their mind, and the truth almost never emerged from the clash of opinions.

The first scientific confirmations of Democritus's suggestion appeared at the beginning of the nineteenth century. By that time, it had been established that any material could be broken down into some basic chemical elements, such as iron or oxygen. The English scientist John Dalton studied these materials and found that they reacted with one another in a very peculiar way. For example, 100 grams of carbon may react either with 133 grams of oxygen and produce carbon monoxide or with 266 grams of oxygen and produce carbon dioxide. Neither Dalton nor anybody else could observe the reaction of 100 grams of carbon with, say, 400 grams of oxygen. To understand such behavior, Dalton recalled the ancient hypothesis regarding atoms. If there is an atom of carbon and an atom of oxygen, which is 1.33 times heavier, then the measured proportions will just mean that the atom of carbon reacts with either one or two atoms of oxygen. This reasoning seemed very convincing, and no other assumption could explain the experimental observation equally well. With time, the atomic model continued to receive other confirmations from various studies and finally became the dominant scientific theory. Democritus could celebrate a well-deserved victory.

But science never stopped in its development. By the beginning of the twentieth century, it was found that atoms were not indivisible, contrary to their name. They could be separated into parts. First, particles called *electrons* were identified. They formed the shell of an atom. After that, a *nucleus* in the center of the atom was discovered.

In a simplistic form, the electrons can be thought of as orbiting a nucleus the way planets revolve around the Sun. This presentation is very far from reality but is nevertheless often used to picture the atom. The nucleus is very small relative to the atom. If the atom were magnified to the size of a football field, the nucleus would have the size of a tennis ball in the center of this field, and the electrons would be somewhere near the spectator stands.

An electron has an important property called *electric charge*. The value of charge always remains the same. By convention, the charge of an electron is negative, while a nucleus is positively charged. Any charged object exerts an *electric* or *magnetic* force on other charged objects. In particular, the electric force keeps the nucleus and the electrons together in the atom.

An electron was discovered because of its charge. The flow of electrons was first observed in the second half of the 19th century as an unusual glow emitted by a negatively charged plate placed in a vacuum. This experiment was quite simple to carry out, but understanding the observed phenomenon was much more difficult. It took more than twenty years to find the right answer. First, scientists observed that the glow casts a shadow. Since light has no shadow (see Fig. 3.1), they concluded that some material particles produced the glow. Later, it was found that these particles have charge. Finally,

Fig. 3.1   The simplest experiment in particle physics: light indeed has no shadow.

in 1896, the English physicist Joseph John Thomson measured their charge and mass and thus proved that the particles are distinct from all known atoms. The new particles were called electrons.

The nucleus, in its turn, was discovered in 1911 by the British scientist Ernest Rutherford. He made this discovery by analyzing the results of research performed by Hans Geiger and Ernest Marsden. Their experiment can be compared to baggage screening at airports using X-rays.[1] The X-rays easily pass through clothes packed in the bags, while more dense materials, like metals, deflect them. The deflection casts a shadow of the dense objects in the beam of X-rays. This beam with shadows in it is transformed into a visible image, which is projected on a screen of a monitor and is scanned by the security staff.

Geiger and Marsden attempted to uncover the inner structure of the atom in a similar way. They directed energetic particles emitted by radioactive material onto a thin foil of metal. The contemporary model of the atom presented it as a cloud of uniform positively charged substance. The electrons were thought to be spread within its volume like plums in a pudding. In such a model, the energetic particles should pass through the substance with almost no deflection. Indeed, this was often the case. But sometimes the foil deflected particles at very high angles. It even reflected them back. This result was entirely impossible within the plum-pudding model of the atom. The researchers themselves were greatly surprised. Rutherford noticed: "it was almost as incredible as if you fired a 15-inch shell at a piece of tissue paper and it came back and hit you."

Thus, the Geiger-Marsden experiment completely disproved the plum-pudding model of the atom. Instead, Rutherford hypothesized that in the atom there must be a small but very dense nucleus containing almost all the atom's mass. Only such a heavy, dense nucleus could cast back the energetic particles that hit it. Conversely, the particles missing the nucleus would pass through the foil with almost no deflection. Rutherford's idea has turned out to be true.

---

[1] The Geiger-Marsden experiment was carried out, of course, much earlier than any baggage screening.

Soon after the discovery of the nucleus, scientists established that it is also a composite object and that it consists of smaller "bricks" called *protons* and *neutrons*. These particles, and several other similar objects are collectively called *baryons*. The charge of a proton is positive. The absolute value of the charge of protons and electrons is the same. The neutron, however, is a *neutral particle*, that is, its charge is equal to zero.

The comparison of baryons to bricks is appropriate. The nucleus of any element can be obtained by a simple combination of a certain number of protons and neutrons, the way a model of a spacecraft or ship can be built from Lego blocks. For example, the simplest nucleus of *hydrogen* consists of one proton while the nucleus of *helium* contains two protons and two neutrons. The largest known nucleus is made of more than 100 protons and around 150 neutrons. New elements, even those that do not exist in nature, can be obtained by adding protons and neutrons to different nuclei.

The transmutation of one element into another was the dream of alchemy, the ancient precursor of modern science. More precisely, alchemy was interested in the synthesis of only one particular element. For many centuries alchemists spent their entire lives trying to find the philosopher's stone that could turn iron, or any other metal, into gold. As a by-product of their search, alchemists made many important discoveries in chemistry and established many properties of materials. Thus, their efforts were not entirely useless. For example, the Hamburg alchemist Hennig Brand discovered phosphorus in 1669 while trying to extract gold from urine. He selected such an unusual object of study because its color was reminiscent of the precious metal. The actual procedure of his research was quite disgusting, but Brand selflessly followed it (or perhaps he just wanted to become rich). In one way or another, an ample dose of imagination, scientific dedication, down-to-earth interests, or an interplay of all these factors resulted in the discovery of the first element since ancient times. This great result, of course, is valued much more than gold today. It immortalizes the name of Brand. But the primary goal of alchemy, the philosopher's stone, was never achieved, not by Brand, nor by anybody else.

From the heights of contemporary knowledge of atomic structure, the idea of turning iron into gold does not seem too crazy. Here is a recipe. Take three nuclei of iron, add one proton and 28 neutrons (count them carefully!), thoroughly mix everything, and you will get genuine pure gold, which you can even trade on stock exchange. The only inconvenience is that the cost of this method exceeds the possible gain by many thousand times. Thus, the industrial synthesis of gold is still reserved for science fiction. But work in this direction continues.

For some time the proton and neutron bore the honorable title of *elementary* (read: indivisible) particles. But less than forty years after their discovery, scientists found, with amazement, that these particles were themselves made of other objects called *quarks*.

3.2   Quarks and Leptons

The inner structure of protons was uncovered using a method not dissimilar to the Geiger-Marsden experiment. However, the energy of particles that could penetrate inside a proton needed to be much higher. Radioactive materials could not produce such energetic particles. Therefore, the discovery of quarks only became possible when physicists learned to build special instruments called *accelerators*.

Simply speaking, an accelerator is a sealed pipe with almost all the air pumped out from inside it. This pipe can be bent into a ring or kept straight. Particles, such as electrons or protons, are accelerated inside the pipe and gain a very high kinetic energy. The role of these particles in accelerator experiments is similar to the task of artillery shells bombarding an enemy castle. From a physical point of view, a shell is a carrier of energy which is used to destroy castle walls. Something similar happens in accelerator experiments. Accelerated particles are directed onto a *target* made from some particular material and strike its atoms. As a result of the collision, the kinetic energy of the particles accumulated during acceleration transforms into radiation. This radiation penetrates inside the particles that form the atoms of the target and reveals their structure.

Quarks were discovered at the Stanford linear accelerator (SLAC) in the USA in 1968. This facility accelerated electrons. In such a process, the electrons gain high energy and collide with protons, which are always present in any material. The researchers studied the electron scattering after collisions. They observed that the electrons were often deflected at high angles, which would not be possible if the proton did not have any structure. This effect was somewhat similar to that observed in the Geiger-Marsden experiment where the energetic particles emitted by the radioactive source were cast back by nuclei. Following this observation, the results of the SLAC experiment proved that the proton is not a uniform substance and is made of smaller objects.

In fact, by the time of the SLAC experiment, the American physicists Murray Gell-Mann and George Zweig already predicted the observed effect. They worked independently but put forward the same hypothesis. They supposed that baryons and some other particles are not elementary and are made of smaller objects. Gell-Mann named them *quarks*. The results of the SLAC experiment agreed very well with the prediction of the *quark model* and, hence, confirmed its validity. Since then, this model has been considered to be the correct theory that describes the inner structure of particles. Consequently, quarks have been treated as the next level in the structure of matter.

Thus, the structure of matter, as we now know it, is very similar to Matryoshka, a set of dolls nestled one inside another. Four layers in this structure (atom, nucleus, baryon, and quark) have been discovered so far. Each of them is concealed beneath the upper layer like the doll-daughter is hidden inside the doll-mother. However, contrary to the dolls, which all look alike except for their change in size, each new layer of the matter is completely different from the previous. Each one has its distinctive features, and, therefore, its discovery entirely changes our understanding of nature. By unearthing them one by one, we reveal new, previously hidden secrets of the universe. After all, this result is ultimately the main goal of scientific research.

A vertiginous shrinking of time intervals separating the discovery of each new layer in the structure of matter, from the misty idea of atoms to their discovery, from atoms to baryons and then from baryons to quarks and electrons, clearly demonstrates a fast acceleration of scientific progress. At the dawn of civilization, advances were made so slowly that they were akin to a snail on the slope of a mountain,[2] but they gradually increased pace to finally achieve breakneck speed. Judging by this quick development of science, it could be expected that scientists will very soon unveil the next level of matter and will find out what is inside the quark and electron.

But the fission of matter has suddenly stopped at the level of quarks. It is not because there have been no attempts to penetrate even deeper inside. Researchers continuously strove to do it in the past and have not ceased their attempts in the present. However, despite the tremendous efforts of an enormous number of scientists and considerable resources spent on their research, nothing more elementary than the quark and electron has been discovered yet. Does this mean that the indivisible particles of Democritus are finally found? This is still a profound mystery. But each mystery is always confronted by the perseverance of people and their tireless endeavor to understand the world around them. Therefore, the efforts to uncover the next levels in the structure of matter will never end. Each new summit conquered in this search leads invariably to a new understanding of the forces of nature and their submission to the needs of the humanity. It was so in the past; it will be the same in the future.

Even though the inner structure of the quark and electron has not yet been found, much has been done to understand their properties and much has been achieved in this direction. According to the quark model, each baryon, like the proton and neutron, contains three quarks. The quarks can be of several different types. The proton is composed of two quarks of type *up* and one quark of type *down*. The neutron in its turn is made of two down quarks and one up quark. In addition to the up and down quarks, four other species of quarks have been discovered. They are not contained in ordinary matter and can be produced only in accelerators. Nevertheless, these additional quarks are essential constituents of matter. They define properties of the universe and are necessary for its existence in the current form.

---

[2] An allusion to the haiku of Kobayashi Issa (the Japanese poet and Buddhist priest): "O snail, climb Mount Fuji but slowly, slowly!".

All six types of quarks belong to the same level of the structure of matter, like siblings in a family. They have quite funny names, rather reminiscent of the names of the dwarfs in the fairy tale of Snow White. The up and down quarks are followed by the *charm*, *strange*, *top* and *bottom* quarks. The last quark is also called *beautiful* or *beauty*. Judging by this choice of names, physicists seemingly had become tired of unpronounceable names like "nucleoside triphosphate" and decided to use normal words that could be understood by all people. Another peculiar feature: physicists say "the flavor of quark" instead of "the type of quark", although, of course, quarks have no flavor at all. In this book, we shall follow this convention and use the term *flavor* to describe different types of particles.

Two particles similar to the electron have also been found. They are called *muon* and *tau lepton*. This latter particle is also called *tauon*. In the name "tau lepton" the first part can be considered as a forename corresponding to the Greek letter τ while the word "lepton" is similar to a family name. This term originates from the Greek word meaning "small" or "light." In fact, the tauon is not so light. It is much heavier than many quarks. It gets its name because of its direct relationship with the electron and muon. Initially, only the electron was known. It is indeed a very light particle. After that, a muon was discovered. The muon is much heavier than the electron. But it is still light compared to the baryons. All properties of the muon, except its mass, coincide with that of the electron. Therefore, it is quite natural to use a common name to denote these two particles. The name chosen was *lepton*. Later, a tauon was found. This particle should by no means be treated as small or light. However, if the muon and electron became as heavy as the tauon, nobody would be able to distinguish them. Because of this similarity, the electron, muon, and tauon all comprise the group of leptons, and that is why the tauon has its contradictory name.

In addition to these three particles, the most imperceptible particles, called *neutrinos*, are also included in the ranks of leptons. These particles fill all space around us, yet for a very long time, nobody knew of their existence. When we walk, sit, sleep or simply lie on a sunny beach, 65 billion (65,000,000,000,000) neutrinos pass through each square centimeter of our skin every second. Ponder this: each second and through each square centimeter! The mere thought of this invisible flow could spoil any peaceful holiday at the most fashionable resort. But don't worry; the neutrinos don't harm us. They almost don't interact with the particles of our body and therefore don't produce any effect. The meaning of the word "interaction" in the context of particles is explained later in this book. Here we will just say that it is precisely because of its weak interaction with other particles that the neutrino can pass through the whole of Earth without hitting any atom on its path. No other particle can do this. The neutrino is imperceptible and difficult to detect, particularly because of this property. Still, scientists know how to catch the neutrinos and study them. They are even able to distinguish three flavors of these particles. These flavors are called *electron neutrino, muon neutrino,* and *tauon neutrino*. Such terms stress the direct relation between the given flavor of neutrino and the corresponding lepton.

Since all neutrinos are counted as leptons, the total number of lepton flavors is equal to six and is the same as that of quarks. Although quarks and leptons are quite different, some of their properties are similar. Whenever only this similarity is important, the quarks and leptons are united under the common name of *fundamental fermions*. The term fermion honors the Italian physicist Enrico Fermi, who made a significant contribution to particle physics.

One of the main properties, which unites all quarks and leptons and at the same time makes them different, is called *mass*. In physics, this term denotes the amount of matter contained in a given object. That is why the fundamental fermions are also called *particles of matter*. The mass of all particles of the same flavor is the same and never changes. Thus, it is the main quality that distinguishes one particle flavor from another.

The neutrinos have the smallest mass among all fermions. For a long time, they were supposed to be massless particles. But their non-zero mass has recently been firmly proven. Moreover, the masses of all three neutrino flavors are different. However, these masses are so small that their actual values have not yet been measured. The masses of all other particles are well known. They differ considerably

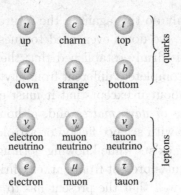

Fig. 3.2   Names and notations of quarks and leptons.

for various flavors. To put this in perspective, if the mass of an electron were the same as the mass of a ping-pong ball, the heaviest known fermion, the top quark, would have the mass of a sports car.

The names and notations of all quarks and leptons are collected in Fig. 3.2. According to current understanding, the six quarks and six leptons presented in this figure constitute the building blocks of matter. All these particles are considered indivisible by modern science because no inner structure inside them has been found. That is why they are called fundamental or elementary. Nobody would suppose that a sports car would have the same (lack of) complexity as a ping-pong ball. Hence, it is hard to believe that the extremely heavy top quark is as elementary as the electron. Nevertheless, this is the current state of particle physics. An extension of the list of elementary particles has also stalled. Despite numerous attempts to find new quarks and leptons, their count is frozen at the magical number twelve. Moreover, many results — both experimental and theoretical — put doubt in the possibility of any further increase in this number.

But this does not mean that new forms of matter will never be found. "Never say never" is a golden rule not only for spies,[3] but also for scientists. Nature has very often presented us surprises in the past,

---

[3]The actual rule of spies is "Never say never again." It is the title of one of the James Bond movies.

and it could certainly do this again in the future. In the case of the structure of matter, we don't even need to guess if there will be a surprise or not. It is firmly established that there is a special type of matter which is completely different from anything else. However, nothing is known about it except that it must exist. It is called by the intriguing name of *dark matter* and, although mysterious, has nothing in common with science fiction or astrology. How can we be sure that something exists while knowing nothing about it? It is like waiting for a Christmas present from Father Christmas in childhood. We know perfectly well that the present has to appear beneath the Christmas tree, but exactly what Father Christmas has brought us will be revealed only on Christmas morning when we rush to the tree, search for our present beneath it and tear up the lovely but unnecessary wrapping paper.

3.3   Dark Matter

Knowledge of the existence of dark matter came to us from the stars. It follows from an observation of the unexpected behavior of galaxies, which would not obey the laws of physics without this additional type of matter. Swiss astronomer Fritz Zwicky was the first to realize this in 1933. He studied the rotation of galaxies around their common center in a giant agglomerate called the Coma Cluster. It turns out that the universe contains not only stars bound together in galaxies, but also galaxies gathered in huge clusters. Zwicky explored one of such clusters containing more than a thousand galaxies.

Analyzing his results, Zwicky found that galaxies near the edge of the Coma Cluster rotate around its center much faster than it is allowed by physical laws. The orbital motion of any object, such as that of a satellite around the Earth, the Earth around the Sun, or the galaxy around the center of the Coma Cluster, requires a force which attracts the object to the center and does not allow it to escape into the depths of space. In all cases, this force is a gravitational attraction between two massive bodies, as first established by Newton. A faster moving object requires a larger force. The attraction of the Sun decreases with the increase in distance to it. Due to this reason, planets located at the edge of the solar system have lower speed than those that are close to the Sun. For example, Uranus orbits the Sun with the speed of about 7 km/s, while the rotational speed of the Earth is 30 km/s. If Uranus at its current location moved with the speed of the Earth, it would immediately break out of the embrace of the Sun's gravitational attraction and would dash away into the infinite vastness of space.

Zwicky measured the speed of the galaxies far from the center of the Coma Cluster, computed the force of attraction produced by all visible matter of the cluster and compared these two quantities. They did not match. The attractive force was much smaller than the value required for the orbital motion of the galaxies with their observed speed. They could not be kept in the Cluster by the force computed. From this observation, Zwicky came to the only possible conclusion. The Coma Cluster must contain some other matter of a special kind. Ordinary matter consists of atoms. Because of this, it is called *atomic matter*. This type of matter undergoes nuclear reactions inside stars and, as a consequence, emits light. Thus, the amount of atomic matter in the galaxies can be determined from their brightness. Apparently, the special kind of matter does not follow the laws of the atomic one. It does not form the stars and does not emit any light. Therefore, it remains invisible to us. For this reason, it is called *dark matter*. Nevertheless, it attracts the visible galaxies by the gravitational force produced by its mass. This additional force keeps the speedy galaxies within the Coma Cluster and does not allow them to escape. The amount of dark matter in the Coma Cluster must be huge. According to the latest calculations, it should exceed the atomic matter of the cluster by ten times. Only in that case can the motion of the galaxies be reconciled with physical laws.

Zwicky's observations were not appreciated in the scientific community and remained largely unnoticed until 1975 when a discovery was announced by American astronomers Vera Rubin and Kent Ford at a meeting of the American Astronomical Society. By that time, highly precise instruments were constructed, allowing scientists to not only distinguish separate stars in galaxies but also to measure the speed of their rotation around the center of the galaxy. By using such devices, Rubin and her colleagues found that the speed of stars is too large for them to be retained in the galaxies by the mass of atomic matter alone. From this observation, the astronomers repeated the statement made by Zwicky about the presence of non-luminous (that is dark) matter in the galaxies.

And so at that time, the idea of dark matter, as well as its name, was accepted by scientists. In fact, the first observation of discrepancy between the observed and expected orbital speed of stars in galaxies was reported by the American astronomer Horace W. Babcock in 1939, in his Ph.D. thesis. But he did not make a conclusion on the presence of additional matter. Instead, he explained his result with the influence of other indirect phenomena. It seems that a lack of imagination can sometimes prevent scientists from making important discoveries.

Thus, by studying stars and galaxies, we can be confident that dark matter exists. Moreover, it constitutes the largest part of all matter in the Universe. According to the most precise measurements, the amount of dark matter is about five times greater than that of ordinary matter. This number is quoted as the average abundance of dark matter in the universe, but there are some galaxies that are made almost entirely of dark matter. Although its amount is huge, we don't know at all what objects constitute dark matter. Are they new fermions or something completely different? We can only guess. There are numerous hypotheses on the nature of dark matter, but none of them have been confirmed experimentally yet. Obviously, there is not much dark matter on Earth and in its vicinity, and there are good reasons for this. Still, the search for dark matter continues at the accelerators and in the mines, at the international space station and in the distant galaxies. There is, therefore, hope that one day, the particles which constitute dark matter will be found. Hope is the last to die.

3.4   Antiparticles

According to our current knowledge, all matter in the universe is made of either quarks and leptons or dark matter (see Fig. 3.3). In addition to these forms of matter, there is also a special group of

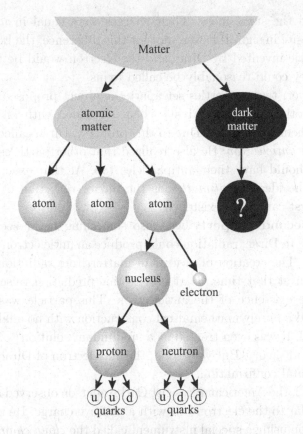

Fig. 3.3   Structure of matter.

objects, which mediate the interactions between all other particles. The representatives of this group and the corresponding interactions will be discussed in the following chapters. Still, even this extended list is not complete. Each fermion has its almost exact copy called the *antiparticle*.

The study of antiparticles began in 1928 when the English physicist Paul Dirac, who was just 26 at that time, formulated a theory describing the behavior of electrons. The theory correctly explained many properties of these particles. It greatly extended the general understanding of other fermions. But it also had a special feature. It predicted the existence of a very unusual object. In many respects, the predicted particle was like the electron. In particular, both of them had the same mass. Their charges were equal in magnitude and opposite in sign. If it were not for this difference, the behavior of the particle invented by Dirac and the electron would be the same. Thus, they could reasonably be called twins.

However, two properties set apart the object proposed by Dirac from all other particles. **First**, when it collided with an electron, both of them *annihilated*. Due to this property, Dirac called the new object an *antielectron*. He also realized that other particles, like the proton, should have their antiparticles too. At that exact moment in time the idea of *antimatter* as the mirror reflection of ordinary matter first came into existence.

The **second property** was also something never seen before. According to Dirac, radiation could produce an antielectron and electron pair. The creation of objects of matter from radiation was not yet known at that time. That is why this prediction raised doubts about the existence of the antielectron. This particle was initially considered a purely mathematical construction with no analog in the real world. It was even treated as a "redundant solution" of the otherwise good theory. But very soon, the antielectron of Dirac received experimental confirmation.

In 1932, the American physicist Carl Anderson observed a particle very similar to the electron but with a positive charge. He made this observation using a special instrument called the *cloud chamber*. This device was invented by the Scottish scientist Charles Wilson. It was a hollow and tightly sealed cylinder with a window on one side. The internal volume of the cylinder was filled with a *supersaturated* water vapor. The vapor in this state was very close to condensing and was ready to turn into a mist of minute water droplets. Any charged particle that crossed the cloud chamber could cause such a mist. However, the mist was formed only near the particle trajectory. As a result, the passage of a particle through the cloud chamber left a clearly visible *track* in its volume. A plane flying high in the sky produces a similar contrail.

The experiments involving the cloud chamber consisted of photographing the tracks appearing inside its volume. Using these photographs, scientists tried to understand the properties and behavior of particles. This task was hard. It was like attempting to decipher the structure of a plane from its contrail. In addition to many other difficulties, the cloud chamber reacted only to charged particles. Neutral particles crossed it without any trace. Therefore, the pictures produced by the cloud chamber were far too incomplete. Still, studying them, scientists succeeded in making many significant discoveries about the microworld of particles. Wilson, in his turn, received a Nobel Prize for the invention of the cloud chamber. This award confirmed the importance of the instrument.

Anderson used the cloud chamber to study *cosmic rays*, which were the main source of information about particles before the advent of accelerators. Cosmic rays are produced by the flows of highly energetic particles, which come to Earth from the depths of space. These rays collide with molecules of air in the upper layers of the atmosphere. The collisions generate showers of secondary particles, which continue to carry all the energy of the original guests from space. Some of these secondary particles reached Anderson's cloud chamber and left trails there.

The setup of Anderson's experiment included a strong magnet located near the cloud chamber. A *magnetic force* produced by the magnet acted upon moving charged particles and deflected them so that their trajectories deviated from a straight line into an arc. The curvature of the arc depended on the kinetic energy of the particle. Higher energy resulted in a smaller deflection from the straight line. Therefore, it was possible to judge the energy of particles by measuring the track curvature. Besides, positively and negatively charged particles were deflected in opposite directions. This effect allowed Anderson to find out the particle's charge. Apart from the magnet, Anderson included one more important piece in his setup. He placed a lead plate inside the cloud chamber parallel to its axis. This addition played a crucial role in the discovery of the new particle.

Anderson analyzed more than 1300 shots of different events that occurred inside the cloud chamber. Among the collected pictures, he found a photograph reproduced in Fig. 3.4. It displayed a very unusual event. A particle crossed the volume of the cloud chamber and produced a track. This track can be seen in Fig. 3.4 as a thin

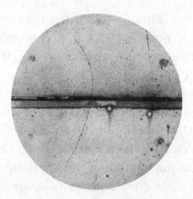

Fig. 3.4   Track of the positron in the Anderson's experiment. (Taken from: Anderson, Carl D. (1933). "The Positive Electron." *Physical Review* **43** (6), 491).

curved line in the center. Anderson very quickly realized that the particle must be similar to an electron. This decision was based on the *path length* of the particle. Any particle passing through a medium collides with its atoms, slows down and eventually stops somewhere inside its volume. The average flight distance mainly depends on the particle type and its kinetic energy. If the particle were a proton with energy corresponding to the observed track curvature, it would travel no more than 5 mm. The actual length of the track exceeded 10 cm. Only an electron could fly such a long distance. The interpretation of the observed particle as an electron would be perfectly fine, but there was a "small" problem. The electron has a negative charge, while the observed particle was positively charged.

The charge of the particle was worked out by using the magnet and the lead plate in the center of the chamber. This plate is seen in the picture as a thick dark strip dividing the chamber into two parts. In Anderson's experiment, magnetic force deflected all negatively charged particles to the right with respect to their direction of motion, while positive particles were deflected to the left. Therefore, if the direction of the particle's motion is known, its charge will be firmly established. But how could the direction be determined from the obtained photograph? The particle could move both upward and downward. Here came the help of the lead plate. It can be clearly seen from the picture that the track was much more curved in the upper part of the chamber than at the bottom. This feature indicates that the kinetic energy of the particle had been higher at the bottom. The particle could lose a part of its energy after crossing the lead plate because of interactions with its atoms. In such a case, the direction of the particle must be upward. The opposite direction would imply that the particle gained some energy after crossing the plate, which is never possible. Based on this consideration, Anderson concluded that the particle was deflected by the magnet to the left, and, consequently, had a positive charge. Since in all other respects the observed object was similar to the electron, it could mean just one thing. Anderson discovered a new particle, the positive electron.[4]

Anderson named his particle *positron.* But it was clear even at the time of discovery that the particle looked like the antielectron predicted by Dirac a few years before. However, "looks like" is never equal to "is the same" in science. Thus, the correspondence of the theoretical particle and the one Anderson discovered had to be proven.

As stated above, Dirac's antielectron has two important properties, which make it an extraordinary object in the microworld of particles. Therefore, the identity of the antielectron and positron can be established only if and when the positron displays these two properties. The conversion of radiation into an electron and positron pair was observed by the English physicist Patrick Blackett soon after Anderson announced his discovery. Blackett was also the first to treat the positron as the antielectron of Dirac. Slightly later, the annihilation of electron and positron was also observed. This finding completed the proof of the antielectron's existence. The astonishment of the scientific community following this discovery was so great that Dirac, Anderson, and Blackett were awarded the Nobel prizes in 1933, 1936 and 1948, respectively.

With time, the corresponding *antiquarks* and *antileptons* have been found for every quark and lepton. Each particle and its antiparticle are very similar. In particular, their interaction with other particles is essentially the same, and their masses are equal.

---

[4] Anderson himself said many times that his discovery was incidental. But nothing happens randomly in science.

They differ only in having opposite charges and some other special properties. The composite particles, such as the proton and neutron, have their antiparticles too. The *antiproton* has been discovered by the American physicists Emilio Segre and Owen Chamberlain. Both of them have been awarded the Nobel Prize. Conversely, the discovery of the *antineutron* made one year later was not similarly rewarded. It seems that by that time the existence of antiparticles had turned from amazing miracles to ordinary reality.

Despite the lack of Nobel prizes, the study of antiparticles actively continued. Soon, the simplest *antinuclei* were produced in accelerator experiments. After that, in 1995, the first *antiatom* of hydrogen, or *antihydrogen*, containing the antiproton and positron was produced. Today, physicists have no doubts that antiatoms can be combined into *antimatter*. Still, no significant amount of antimatter has been produced as yet.

It can be seen that the discovery of different layers of antimatter occurs in the order opposite to that of ordinary matter. First, the antilepton (that is, the positron) was discovered, and after that, the antibaryon, antinucleus, and antiatom were found. Currently, the antiatoms are the subject of extensive study. Several such experiments are conducted at the European Center for Nuclear Research (CERN). In all cases, the production of antiparticles is not as difficult as maintaining their existence. For example, tens of thousands of hydrogen antiatoms were created in the experiment ATHENA. They were produced in a vacuum volume of the experimental apparatus. In principle, they could exist there forever, since the antiatoms are as stable as ordinary atoms. However, all of them moved with high speed, unavoidably collided with the walls limiting the volume, and were annihilated very quickly after their production.

The goal of another experiment, ALPHA, is to store the antiatoms of hydrogen for a long time. This storage should allow their thorough exploration. One of the latest achievements of this research is the trapping of 309 antiatoms for 1000 seconds. Such length of storage opens the possibility of precisely measuring the properties of antihydrogen and comparing them with that of ordinary hydrogen. The properties of an atom and its antiatom should be almost the

same. However, the observation of even a tiny difference can significantly influence our understanding of nature. In one of the studies,
the researchers measure the *spectrum of light* emitted by the antihydrogen atom. Atoms emit or absorb light of specific colors, which
form the atoms' spectrum. The spectrum of each element is unique
and well known. This information permits scientists to determine
the chemical composition of different substances just by observing
their emitted light. In particular, the content of stars is deduced this
way. According to theoretical expectations, the spectrums of hydrogen and antihydrogen should be the same. However, this expectation
needs to be verified experimentally. The first ALPHA results indeed
show no difference between the two spectra. Still, scientists expect to
continue this study and improve the precision of their measurements
with the hope of finding some deviation from theoretical prediction.

The significance of the discovery of antimatter cannot be under-
estimated. It is not just an observation of one more object of the
microworld. This discovery is a major step towards explaining the
origin of matter during the evolution of the universe. According to
the Dirac theory, any particle can be created by radiation. However, the theory also states that the appearance of any particle in
this world is always accompanied by the corresponding antiparticle.
In terms of the previously mentioned riddle about the camel and
the eye of a needle, the discovery of antimatter clearly tells us that
transforming radiation into a camel is not possible without a simultaneous generation of anti-camel somewhere nearby. It turns out that
this conclusion has very far-reaching consequences.

The mechanisms of creation and annihilation of matter are considered in the following chapters. But before doing this, one more
participant of all such processes should be introduced. Previously,
we used the term "interaction" on several occasions without ade-
quately explaining its meaning. Now it's time to discuss it in more
detail. This discussion starts in the next chapter, where the best
known *electromagnetic interaction* is considered. This example will
help us later to understand all other types of interaction, which play
an essential role in the creation of matter and the formation of the
universe.

Chapter 4

Electromagnetic Interaction

4.1 Photon

The description of accelerator experiments in the previous chapter suggests that accelerated particles collide with atoms of a target. Strictly speaking, collisions do not exist in the microworld. All processes there are called *interactions*. Interactions are responsible for everything that happens with particles. The change in any particle's property, such as its energy or direction of motion, occurs only because of interaction with other particles. Interactions also determine the creation and destruction of particles. Therefore, the understanding of particles is impossible without exploring interactions between them.

Currently, five different types of interaction are known. One of them, named the *Higgs interaction* after the Scottish theorist Peter Higgs, was discovered very recently. Scientists know very little about it and spend a lot of time and money on its study. These investigations are performed at CERN, at the most energetic accelerator ever built. Conversely, *electromagnetic interaction* has already been extensively explored and is widely used in many practical applications.

Electromagnetism, as suggested by its name, is made up of two parts, electricity and magnetism. Separately, people have known about both these phenomena for a long time. But their fusion into a single entity has happened relatively recently, while the thorough understanding of electromagnetism has become possible only after studying the properties of particles. Thales of Miletus, one of the Seven Sages of Ancient Greece who lived between the 7th and 6th centuries BC, was the first who attempted to explain the nature of electricity and magnetism. Like all other sages of his time, he did it by using just the power of his mind. Thales had the bright thought that these phenomena are of a common origin. But his idea was successfully forgotten for a long time. To reinvent it, generations of scientists accumulated knowledge about electricity and magnetism during the next two and half thousand years. Philosophical deliberations were complemented by physical experiments and theoretical analysis. Gradually, the collected information added up into a complete picture. The multimillennial endeavor culminated in a system of equations formulated by the Scottish scientist James Clerk Maxwell. This system described in mathematical language all electric and magnetic phenomena, which turned out to be tightly related to each other. Moreover, Maxwell proved that both electricity and magnetism are just two parts of one whole, which was called, for lack of a better word, electromagnetism.

In Maxwell's time, nothing was yet known about elementary particles. Therefore, Maxwell did not include them in his equations. The role of particles has become apparent only after the development of theories explaining their behavior. Now it is firmly established that **interacting charged particles produce all electromagnetic effects**. Each atom contains positively charged protons and negatively charged electrons. It turns out that just these two invisible "gods" and the electromagnetic interaction between them are responsible for most natural phenomena of the world. They cause the flashes of lightning in the sky, St. Elmo's fire on the masts of sailing ships,[1] the glow of the Northern and Southern Lights during long polar nights, and many other spectacular manifestations, which had initially been attributed to magical forces from the beyond.

---

[1]Sailors crossing the seas often observed a bright blue glow on the masts of their ships, which occurred during violent thunderstorms. This light was called St. Elmo's fire in honor of St. Elmo, the protector of sailors. Its appearance was considered a good omen announcing the near end of a mortal peril.

In addition to charged particles, each electromagnetic interaction involves one or several other objects called *photons*. We previously said that all leptons and quarks are fermions. A photon belongs to another class of particles called *fundamental bosons*. In general, each particle can be either a boson or fermion, and this division represents the most general classification of all objects of the microworld. The term "boson" is derived from the name of the Indian scientist Satyendra Nath Bose, who developed a theoretical description of this class of particles. Fundamental bosons play a special role in the microworld. They mediate interactions between other particles. Each type of boson conducts a specific type of interaction. Photons are responsible for electromagnetic interactions.

The origin of the term "photon" comes from the Greek word meaning "light." Such a name is motivated by the fact that light can be treated as a flow of photons. This statement means that photons are everywhere around us. However, nobody knew about their existence for a very long time, up to the beginning of the twentieth century.

To be fair, the idea of presenting light as a flow of particles emerged about three hundred years ago and was even supported by many prominent scientists of that time, like Newton. But later, an opposite hypothesis gradually took the upper hand. In all experiments, the observed behavior of light was very similar to that of an ordinary wave on a water surface. And so, the wave hypothesis of light was developed. The theory of Maxwell confirmed this speculation. Using his equations, Maxwell predicted the existence of *electromagnetic radiation*, which propagated as a wave. He also demonstrated that light was just one form of this radiation. Consequently, light also represented a wave propagating in space. Many experimental results successfully confirmed the conclusions of Maxwell. Thus, by the end of the nineteenth century, nobody treated light as a flow of particles anymore. That is why the first appearance of the photon as a particle of light came as a real shock to the scientific community. This debut deserves separate consideration.

At the end of the nineteenth century, researchers noticed that light shining on some types of materials (mainly metals) dislodged the electrons from them. This phenomenon was called the *photo-electric effect.* At first glance, the photoelectric effect fits well with Maxwell's theory. Electrons are charged particles. If the light is a form of electromagnetic radiation, it can act upon electrons and transfer its energy to them. In the photoelectric effect, this additional energy allows electrons to escape from the material. However, a deeper study revealed that some features of the photoelectric effect were inconsistent with the wave nature of light.

To realize all problems of the photoelectric effect, we can use the analogy between electromagnetic radiation and a conventional wave propagating on the surface of the sea. Evidently, a high ocean wave strikes ships standing in its way with much more force than a less energetic lower wave. Similarly, the wave theory of light implies that higher energy of incident light must transfer a higher amount of energy to the ejected electron. The energy of light is determined by its brightness. This relationship is natural and agrees well with common sense. Indeed, lying on a beach, it is more likely to burn one's skin on a sunny day than on a moonlit night. However, it turns out that in the photoelectric effect, the electrons, which are knocked out by both bright and dim light, receive the same amount of energy. In other words, the energy of the ejected electrons does not depend on the radiation energy of the incident light.

This inconsistency, even if it looks minor and insignificant, is in fact very serious. It is equivalent to the claim that a light sea-swell strikes a ship with the same force as a tsunami. Clearly, this is nonsense. But the photoelectric effect could not be called nonsense because it was experimentally proven. So, the strange behavior of light needed an explanation. The resolution of the controversy was proposed by Einstein. He suggested that electromagnetic radiation, including light, propagates in spatially localized packets which he called "quanta of light." In this interpretation, the photoelectric effect is produced by absorption of a quantum of light by an electron. Such absorption can be compared to the hit of a stationary billiard ball by a bullet. After this hit, the ball receives some part of the bullet's kinetic energy and starts moving. Similarly, the electron absorbs the quantum of energy brought by light and escapes from the material with its help.

There is a significant difference between the strike of a radiation wave and that of a quantum of energy. A radiation wave can be considered continuous in space. Thus, the electron that stands in its way cannot avoid its action. Recalling the analogy of sea waves, a ship can never evade the wave moving towards it. On the contrary, localized portions of energy, like bullets, can hit the electron, but they can also miss it. It turns out that the misses happen much more often than the hits so that each electron has a chance to catch at most only one quantum. If this happens, the electron's energy will depend only on the energy of the single quantum and not on the total number of quanta incident on the material.

The next important step made by Einstein was finding the energy contained in a quantum. He demonstrated that the brightness of light affects the total number of quanta in the beam, but not the energy of each quantum. Using the results of the German theorist Max Plank, Einstein concluded that the energy of the quantum depends only on the color of light. Contrary to naive expectations, the quanta of "cold" blue light are more energetic than that of "hot" red light. However, all quanta in a bright or dim beam of light of particular color have the same energy. Hence, Einstein explained why the energy of escaped electrons is not sensitive to the brightness of the incident light. He also predicted that this energy must depend on the color of light.

The main idea of Einstein was to describe the action of light on electrons as a discrete interaction of the electromagnetic quantum and the particle. This idea completely contradicted the presentation of light as a form of electromagnetic wave. Consequently, many scientists initially did not accept Einstein's hypothesis. The American physicist Robert Millikan was so skeptical that he conducted a dedicated experiment for ten years in an attempt to disprove the idea of Einstein. In particular, he wanted to demonstrate that the color of light does not influence the energy of electrons knocked out in the photoelectric effect. However, his results turned out contrary to his intention; they were in perfect agreement with Einstein's prediction, although Millikan believed for a long time that this coincidence was accidental. In any case, Einstein received the Nobel Prize in 1921 for his explanation of the photoelectric effect, while Millikan was awarded the same prize two years later. And so, the quantum nature of light had been firmly established.

The quanta of light were called photons. Very quickly, theorists realized that the photoelectric effect was not something exceptional. They built a theory, in which any electromagnetic radiation represents the flow of photons, and all electromagnetic interactions consist of the exchange of photons between charged particles. The theory is called *quantum electrodynamics*, or *QED*. In this theory, the process of interaction is reminiscent of a football game. From a physical point of view, this game can be described as a series of separate and clearly defined passes of the ball from one player to another. Hence, the pass can be treated as a *football interaction* of players, while the ball can be considered as an object mediating it. Something similar happens in electromagnetic interactions. Particles exchange photons just as football players pass the ball to one another. Each process of interaction is discrete, that is, clearly bounded in space and time. Charged particles can emit photons, which are absorbed by other charged particles. This sequence of emissions and absorptions constitutes electromagnetic interaction.

A notable example of electromagnetic interaction is the process responsible for our sense of sight. Any object emits photons. More precisely, photons are emitted by charged particles contained in any object. These photons move in all directions, and, in particular, hit our eyes. There, they are absorbed by other charged particles, which transform the energy of photons into signals sent to our brain. This is how we see the world around us.

The photon is the first fundamental boson discovered. Besides it, several other particles of this class are known. We will consider them later, when we introduce corresponding interactions. In contrast to all fundamental fermions, the photon is a massless particle. The absence of the photon's mass is simple to predict. Light is the flow of photons, and only massless particles can move at the speed of light. The speed of all massive particles must be below this threshold.

Another important feature of photons is that they can exist for an infinitely long time. That is why we see distant stars. Their light travels in space for billions of years before hitting our eyes or the sensitive elements of scientific instruments.

Finishing our introduction of the photon, we also need to answer one more three-word question. What is light? It seems that the concept of photons introduced by Einstein completely confuses understanding of this phenomenon. From one side, light can be presented as a continuous electromagnetic wave, which cannot be divided into parts. There are many proofs supporting this treatment. From the other side, light represents the flow of photons, which are discrete portions of energy well-separated in space and time. So, where do waves end and particles begin? The answer to this question is surprisingly simple. Light is both a propagation of a wave and a flow of particles. In some cases, it demonstrates the properties of a wave; in other situations, it behaves like a particle. But both these forms are just two sides of one whole. They cannot be separated from each other. Such a *unity of opposites* should not be unfamiliar to people. After all, there is a thin line between love and hate, is it not? It is also easy to find physical examples of such a unity. There are many objects which can take several forms and nevertheless, remain a single entity. As is well known, water can be either a hard solid, a flowing liquid or a volatile vapor, depending on its temperature. This example can help to clarify the ever-changing nature of light.

4.2 Feynman Diagrams

Electromagnetic interactions can be visually presented in the form of special diagrams. An example of such a diagram is shown in Fig. 4.1. The drawing depicts the electromagnetic interaction of two electrons. A special graphical language that is used in the picture shall be considered in more detail because diagrams similar to that in Fig. 4.1 are very important for understanding the behavior of particles. They are also often used in this book, necessitating their understanding.

Each full broken line in the diagram of Fig. 4.1 represents a "line of life" of the fermion during the action depicted. In this particular

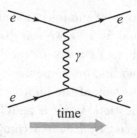

Fig. 4.1   Diagram of the electromagnetic interaction of two electrons.

case, the lines correspond to two electrons, but they can be replaced by any other fermions. The diagram displays the development of the process in time. The time flow is always directed from left to right, as shown in the figure. With some imagination, the electrons can be pictured as small beads moving along their lines in the direction shown by the small arrows. In the initial moment, both electrons are located on the left side of the diagram. This side is called the *initial state* of the process. The electrons start to move along their lines and interact with each other at a certain moment. The interaction consists of the exchange of a photon, which is shown as a wavy line and is denoted by the Greek letter γ.

After the interaction, the properties of both particles change. This is reflected by the break in the electron lines. The diagram does not specify which properties change, nor by what amount. These details are not essential for understanding the process. This is why the angle between the initial and final directions of the electron lines is not significant. What is important is the change of the direction itself. Another important thing is the flavor of the fermion before and after the interaction. In electromagnetic interaction, the flavor of the fermions does not change. However, such a change does occur in some other interactions considered later. That is why the flavor of the fermion before and after the interaction is always indicated in the diagram, even if it remains the same. In Fig. 4.1, the electron is denoted by the symbol e. The notation of all other fermions is given in Fig. 3.2.

When the interaction is finished, the particles continue to move along their lines without any further disturbance. In the end, they come to the right side of the diagram, which is called the *final state* of the process.

The diagram in Fig. 4.1 seems to be no more complicated than a drawing of a six-year-old child. Still, it is full of meaning. It presents the essence of electromagnetic interaction. Pictures like this are called *Feynman diagrams*, named after their inventor, the American physicist Richard Feynman. All five currently known interactions are represented by diagrams similar to the one above, which help physicists to "see" and understand physical processes by displaying them in the pictorial form. Let us also try to determine the main properties of the electromagnetic interaction by just looking at Feynman diagrams.

The diagram of any interaction is constructed from several basic elements called *vertices*. The vertex of the electromagnetic interaction of an electron is shown in Fig. 4.2 (left). Similar vertices exist for the electromagnetic interaction of all other charged fermions. They are obtained by a simple replacement of the symbol of the electron (e) by the symbol of another particle. For example, Fig. 4.2 (right) shows the vertex of the interaction of an up quark (u) and a photon. The vertex of electromagnetic interaction always includes one full broken line corresponding to the fermion and one zig-zag line corresponding to the photon. Two segments of the *fermion line* are called its *legs*. Each leg is marked by a small arrow. We will discuss the meaning of these arrows a little bit later. There is no such arrow on the photon line.

The fermion line in the vertex of the electromagnetic interaction always includes one arrow that "enters" the vertex and another that

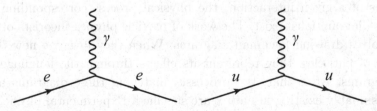

Fig. 4.2   Vertex of the electromagnetic interaction of an electron (left) and of an up quark (right).

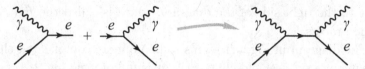

Fig. 4.3   Diagram of photon scattering by electron (right) obtained by combining two electromagnetic vertices (left).

"exits." A diagram in which two arrows enter or exit from the vertex is never possible. In other words, the direction of arrows along the fermion line always remains the same.

Vertices with two or three photons don't exist. However, a diagram containing two photons can be obtained by combining two vertices as shown in Fig. 4.3. In general, to draw the diagram of any process, it is sufficient to take several vertices and join some of their lines. The resulting diagram in Fig. 4.3 (right) corresponds to the scattering of a photon by an electron. The initial state of this process contains the photon and electron. They interact with each other and diverge after that. Their properties in the beginning and at the end are different. For example, the photon can transfer a part of its energy to the electron. This change is reflected by the modified direction of the lines of two particles.

The physical process that corresponds to the diagram in Fig. 4.3 (right) indeed exists. It was discovered in 1923 by the American scientist Arthur Compton. The observation of this process confirmed the quantum theory of light and became an important step towards building the modern theory of particles. That is why Compton was awarded the Nobel Prize for his discovery.

The example of the diagram in Fig. 4.3 illustrates a very important rule. If a diagram is obtained by a combination of several vertices of a given interaction, the physical process corresponding to this diagram must exist. The work of particle physics theorists often involves drawing Feynman diagrams. When they invent a new theory of particles, they represent its effects through the language of diagrams. After that, the processes matching these diagrams and presumably existing in nature are specified. Experimental physicists search for these processes to confirm or disprove theoretical ideas. Such development of particle physics happened many times in the past, and new phenomena predicted by theorists were sometimes indeed discovered. That is why physicists highly value the Feynman diagrams.

The vertex shown in Fig. 4.2 is determined by the properties of electromagnetic interaction. These properties are the following:

First, the photon interacts only with charged particles. Therefore, only charged particles can emit or absorb the photon. Almost all quarks and leptons, except neutrinos, have charge. For each particle, the value of its charge is usually presented relative to the magnitude of the electron charge. In this convention, the charge of the electron, muon and tauon is equal to -1, that of u, c, and t quarks is $+2/3$ and that of d, s, and b quarks is $-1/3$. All these values are collected in Table 4.1.

Since neutrinos are neutral particles, they don't participate in electromagnetic interactions. This is the main reason for their imperceptible nature. The photon does not have a charge either. Therefore, it cannot interact with another photon. In many fantasy stories involving ghosts, the mysterious creatures, so reminiscent of light, can be distinguished from normal people by the absence of their shadow. This unusual feature of ghosts has a clear physical explanation. The shadow essentially means that the photons in a beam of light interact with objects standing in their way. But photons don't interact with other photons, and therefore, they do not have a shadow.[2] Hence, any creature made of light has no shadow either.

Second, electromagnetic interaction always preserves the flavor of the participating particles. This means that an electron can never

Table 4.1. Charges of fundamental fermions

Particle	u, c, t	d, s, b	ν_e, ν_μ, ν_τ	e, μ, τ
Charge	$+\frac{2}{3}$	$-\frac{1}{3}$	0	-1

---

[2]This statement can be verified experimentally. The corresponding test, which can be considered as the simplest experiment in particle physics, is demonstrated in Fig. 3.1.

become a muon. Similarly, a lepton cannot transform into a quark. It should be stressed that this rule is relevant only to electromagnetic interactions. It is violated in some other cases. The consequence of this rule is the **third** important property. Electromagnetic interaction never changes the charge of a fermion. Since the vertices shown in Fig. 4.2 don't modify the charge of the fermion, any combination of vertices also preserves the total charge of particles participating in the corresponding electromagnetic process. This property is generalized as the *law of charge conservation*. It states that the total charge of a group of particles always remains the same regardless of any processes that occur within this group. The individual particles can appear or disappear, but the total charge should never change. This law is as fundamental and unbreakable as the law of energy conservation discussed previously. Not only electromagnetic interactions, but also all others obey it. The law was first established experimentally, but its origin and deep meaning are revealed only after studying the electromagnetic interaction.

The charge of any antiparticle is opposite to that of the corresponding particle so that all antifermions, except antineutrinos, are charged objects. Therefore, antiparticles can also participate in the electromagnetic interaction. For example, the diagram shown in Fig. 4.4 displays the interaction of a positron and an electron. The symbol representing the positron is $\bar{e}$. The bar above the letter indicates that the object is the antiparticle. The same convention is used to denote all other antiparticles. Also, the arrows on the positron line are opposite to the flow of time. This reversal is intentional. It not only distinguishes the particles from antiparticles in the diagrams but

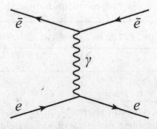

Fig. 4.4   Diagram of the electromagnetic interaction of a positron and electron.

also has a profound physical meaning. It turns out that according to the mathematical description of antiparticles, they can be treated as ordinary particles moving back in time. This may sound completely impossible, but this is the actual outcome of mathematical formulae. As was already discussed, mathematics never hesitates to claim the existence of phenomena which are unimaginable in normal life. Does this mathematical treatment mean that the antiparticles are time travelers coming to us from the future? Attempts at answering this question would be a great subject of a separate book. But we will not elaborate any further here. All that is required to be mentioned for the following discussion is that **whenever the arrows on the fermion lines of Feynman diagrams are directed opposite to the flow of time, these lines correspond to antiparticles**.

The diagram shown in Fig. 4.4 presents the interaction, in which an electron and positron meet but don't destroy each other and continue to exist in the final state. The corresponding process must occur in nature. Thus, just by drawing the Feynman diagram, we can derive a very interesting conclusion. It seems that annihilation is not the only option for the colliding particle and antiparticle. They can just exchange a photon and continue moving on their own way.

Still, destruction and creation happen in the microworld of particles like everywhere else. Such events are also caused by interactions and, hence, are described by the corresponding Feynman diagrams. These processes are essential for understanding the emergence of matter during the universe's evolution. That is why we discuss them in the following chapter.

Chapter 5

Creation and Destruction
of Matter

5.1 Matter and Energy

Many natural phenomena seem like miracles that could never happen in real life. The creation and destruction of matter certainly fit into this category. A few hundred years ago, they could puzzle any philosopher with the richest and most sophisticated imagination. None of the serious scientists expected that generation or annihilation of matter would ever be possible. The discussion of these phenomena would surely appear to be more appropriate in a textbook on magic than in scientific research. Yet today, physicists know not only how to observe and explore these processes, but also how to utilize them for the needs of people.

For many hundreds of years, classical science assumed the conservation of total mass of matter regardless of any transformations of objects containing it. By the end of the eighteenth century, this principle was tested multiple times and was approved by the scientific community as one of the main laws governing nature. The most significant modifications of matter were observed in such processes as transitions of liquids into gases and chemical reactions turning one substance into another. The total mass of all materials was thoroughly measured before and after these processes, and it always remained the same.

However, after the discovery of particles and the first penetrations inside the atom, the classical concept of matter and its conservation started to crack under the pressure of experimental results. Let us consider, for example, the mass of a nucleus made of one proton and one neutron. Such a nucleus is called *deuterium*. The mass of all particles, including the proton and neutron, is extremely small. Measuring it in grams or ounces is inconvenient. That is why the mass of particles is expressed using special units called *GeV*.[1] In these units, the mass of the proton is 0.93827 GeV, while the mass of the neutron is 0.93957 GeV. These values are given here with many digits after the decimal point, but this level of precision is required for the following discussion. Let us try to predict the mass of deuterium. Adding the numbers together results in 1.87784 GeV. However, the actual measurement of the deuterium mass gives a slightly different value of 1.87561 GeV. Thus, the residual between the expected and actual mass is 0.00223 GeV. Of course, this is a very small quantity, but it is still non-zero. There is no mistake here, this difference is firmly established. The combination of the proton and neutron into the nucleus of deuterium does decrease the total mass.

It is pointless trying to argue with numbers. Therefore, the example given above leads to the only possible conclusion. The law of matter conservation is violated in the microworld of particles. After accepting this, it remains just to understand what happens with the excess of matter after the proton and neutron produce the deuterium. It turns out that this excess is converted into energy.

Any massive object possesses a special kind of energy, which is called *rest energy*. This energy was theoretically predicted by Einstein at the beginning of the twentieth century. According to his theory, rest energy is present in any object and is proportional to its mass.[2] Rest energy of any group of objects never changes if the total mass of the group remains the same. But if the total mass decreases, the

---

[1]This unit is the abbreviation of the term "*Giga electron Volt*", which by itself sounds very strange. Nevertheless, it needs to be accepted as one more scientific term used in this book.

[2]More precisely, the famous equation of Einstein states that $E = mc^2$ where m is the mass of a particle and c is the speed of light in vacuum.

corresponding part of rest energy is released and is transformed into other forms. In the case of deuterium, the excess rest energy contained in the proton and neutron is emitted in the form of electromagnetic radiation. This radiation is indeed observed experimentally in total agreement with Einstein's theory.

The example of deuterium proves the possibility of converting matter, or at least part of it, into energy. Indeed, the creation of each nucleus of deuterium causes the disappearance of 0.00223 GeV of matter. Instead of it, an equivalent portion of energy is produced. Another example of such a conversion rises every morning in front of our eyes in the east to later set every evening in the west. The Sun heats our planet and shines brightly in the sky just because of the transformation of part of its rest energy into radiation. A long chain of nuclear reactions taking place inside the Sun fuses four protons into the nucleus of helium. The masses of the initial state (four protons) and the final state (the nucleus of helium) are slightly different. This mass difference converts into energy, which is taken away from the Sun by electromagnetic radiation (mainly in the form of ultraviolet, infrared, and visible light) and neutrinos. Radiation that reaches the Earth leaves all its energy here, while neutrinos give nothing since they pass through our planet without stopping. The net result of these processes is a continuous decrease of the solar mass in exchange for the heating of our and all other planets in the Solar system.

In this world, energy is more valuable than anything else. Without it, no other goods can be fabricated. Every machine, computer or vacuum cleaner needs energy to work and produce results. The more energy a nation has, the more grandiose plans it can accomplish. From the physical point of view, the history of all hitherto existing society is the history of a struggle for energy. At the dawn of civilization, the main sources of energy were slaves, animals, and plants. All of them converted the energy of sunlight into other forms. Those who possessed these sources of energy were rich and prosperous. They could utilize generated energy for various purposes, such as laying roads or constructing buildings. That is why neighboring tribes and nations fought endless wars with one another to seize energy sources. During the development of civilizations that followed, people learned to extract energy from rivers, coal, oil, and gas. This ability allowed them to replace manual labor with the work of much more efficient machines. However, the struggle for energy did not stop for a second, producing more and more suffering and victims. It continues to this day, in the form of both open wars and hidden economic and political methods.

Meanwhile, an unlimited source of energy lies literally under our feet. Any material object, be it hydrogen gas, a stone or a distant star in deep space, is a store of conserved energy tied to its mass. This store is really vast. The Sun is just one example proving what an enormous amount of energy can be obtained by gradually destroying matter. If people discover how to harness the processes that release rest energy, they will be able to satisfy all their energy needs for centuries ahead. The temptation of this possibility is so irresistible that enormous efforts are currently applied to reproduce the nuclear reactions occurring inside the Sun, here on Earth. The practical results of this research are still limited. However, scientists are confident that humankind will one day seize this ultimate source of energy. Maybe then, at last, blood will no longer be traded for oil.

5.2   Annihilation

Having established the important relationship between matter and energy, we can consider annihilation in more detail. We start our discussion from the simplest type of annihilation, in which an electron and positron are converted into radiation. This process is very spectacular, if we could only see it. Two material objects meet at a point in space and suddenly vanish. Instead of them, radiation emerges. This radiation is rather energetic, as it carries away all the rest energy of the electron and positron. Radiation propagates from the point of annihilation not like a shockwave after an explosion, simultaneously expanding in all ways. Instead, it manifests itself as two photons, that is to say, two bundles of energy clearly located in space and each moving in its own well-defined direction. ·

The process of electron-positron annihilation can be presented in the form of the Feynman diagram, which is shown in Fig. 5.1.

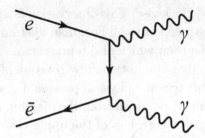

Fig. 5.1   Diagram of annihilation of electron and positron with emission of two photons.

The initial state, that is the group of particles on the left side of the diagram, contains one electron and one positron. The positron is distinguished by the arrow on its line, directed opposite to the flow of time. This flow, in its turn, is usually not displayed on the Feynman diagrams, but is always assumed to be directed from left to right. Annihilation involves two vertices of the electromagnetic interaction. A photon is emitted in each of them. Thus, the final state contains two photons, while the electron and positron, present initially, cease to exist.

The diagram in Fig. 5.1 looks quite different from the actual event of annihilation described above. The distance between two vertices in the diagram is quite large, while in the real event, annihilation occurs at a single spatial point. Also, in the real process, two emitted photons usually move in the opposite directions, while in the diagram their directions are almost parallel. The reason for this difference is that Feynman diagrams never reflect the directions of participating particles, the scale of the event or its duration. Looking for similarities between Feynman diagrams and actual events make as much sense as attempts to find resemblance between some of Picasso's female portraits and their models. Those portraits aim not to show the correct proportions or the shape of a human face, but rather to reflect the inner world of people. The Feynman diagrams, in their turn, present the essence of particle interactions and not their manifestation in reality.

Let us compare the diagrams in Fig. 4.3 (right) and in Fig. 5.1. It is obvious that the second diagram can be obtained from the first by a rotation of 90° clockwise. This observation is a particular case of an essential feature relevant to all Feynman diagrams. Their rotation results in a new diagram, which also represents a valid physical process. Similarly, the diagram obtained by rotation of just one fermion line also corresponds to some physical process. For example, the diagram in Fig. 4.4 can be obtained from the diagram in Fig. 4.1 by the appropriate rotation of both legs of the upper electron line.

The emission of two photons in electron-positron annihilation is confirmed experimentally and is widely used in various practical applications. The most important of these applications is the *positron emission tomography* or PET. This technique is employed in medicine to study the internal organs of the human body. The method of PET consists of the following.

In 1934, just two years after the discovery of the positron by Anderson, the French scientists Irène and Frédéric Joliot-Curie found that some unstable nuclei emit positrons. Ironically, they made photographs of these processes with clearly visible tracks of positrons even before of Anderson's discovery. But they wrongly supposed that the positive particles, which produced these tracks, were protons. Thus, the couple Joliot-Curie missed the opportunity to discover the positron. Yet, the Nobel Prize did not bypass them. They were awarded the prize for other significant achievements.

In the method of PET, nuclei emitting positrons are introduced into the blood of a patient. The blood carries these particles into the part of the body to be investigated. There, the positrons interact with electrons from ordinary atoms and annihilate. The two photons produced in annihilation have high energy. Therefore, they pass unhindered through the body and reach radiation detectors, which surround the patient, as is schematically shown in Fig. 5.2. These detectors can determine the point at which the photon hits, but not the photon trajectory. Nonetheless, the trajectory can be reconstructed, since two emerging photons travel in opposite directions and their trajectories form a single straight line. So, the direction of their motion can be derived from the positions of two opposite hits in the radiation detectors. During a relatively short period of time, millions of annihilation events occur inside the body. Each of them

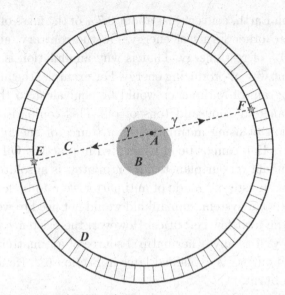

Fig. 5.2   Principal scheme of the measurement of the direction of two photons in the PET method. In this scheme: A — point of electron-positron annihilation; B — the patient body; C — the line of motion of two photons travelling from point A in the opposite directions; D — Detectors of radiation; E, F — positions of the hits of two photons in the radiation detectors.

produces its own photons. The trajectories of photons, which are carefully measured, cross inside the organ of the patient to be investigated. The resulting detection and collection of the crossing points produces a three-dimensional image of the organ, which clearly maps its structure. Such pictures are very informative and are widely used for medical or research purposes.

The production of two photons is a distinctive feature of electron-positron annihilation, but in general, particles can annihilate in many possible ways. All such processes are accompanied by the release of rest energy contained in the particle mass. Since the particle and antiparticle disappear after annihilation, this energy must be transformed into some other form. It can be radiation, as discussed above, but there are also other alternatives.

The amount of energy generated in annihilation is huge. The nuclear reactions that heat the Sun and produce essentially all energy consumed on Earth, convert just about 0.7% of the mass of the initial particles into other forms of energy. On the contrary, annihilation releases 100% of rest energy. That is why annihilation is by far the most efficient way of producing energy. For example, the annihilation of just one gram of antimatter would be equivalent to the burning of more than four thousand tons of oil. This comparison justifies the attraction of using antimatter as a source of energy. Unfortunately, this option cannot be implemented in practice. Oil is still relatively abundant on our planet, but antimatter is absolutely absent. Of course, if an asteroid made of antimatter were to be found somewhere in the solar system, humankind would not need to worry about energy for the next few centuries. However, the existence of such an asteroid, as well as any other natural source of antimatter, has been totally ruled out, as is explained later in Section 6.2. Hence, oil still needs to be burnt.

Nonetheless, at least one application of antimatter as a source of energy is possible in principle. It could be used as fuel for spaceships. The main advantage of this type of fuel is the extremely efficient conversion of mass into energy. Thus, a relatively small amount of antimatter taken on board would be sufficient to propel the spaceship to near-light velocities. However unlikely such an idea seems, this application of antimatter is currently being actively discussed by experts. Moreover, specific designs of an antimatter engine are being developed. The main problem for the realization of this idea is, again, the difficulty in obtaining a source of antimatter. A possible solution is to artificially produce antimatter here on Earth, store it in special containers and load it onto the spaceships. Of course, a big question, which needs to be addressed in this scheme, is how antimatter can be produced. A theoretical answer to this question is already known. It turns out that in the microworld of particles, any process can be inverted in time. In other words, if the conversion of particles and antiparticles into energy is possible, the opposite process — the creation of particles and antiparticles from energy — should also exist. Such transformations are considered in the next section.

5.3   Creation of Matter

Previously, we mentioned that the rotation of the diagram in Fig. 4.3 (right) clockwise results in the diagram of the electron-positron annihilation. The rotation of the same diagram by 90° counter-clockwise produces another diagram shown in Fig. 5.3.

In this diagram, two photons are in the initial state, while the electron and positron are in the final state. Since any diagram obtained by rotation must correspond to some physical process, the creation of the electron and positron by two colliding photons must exist. And it really does. This process can indeed occur, provided the total energy of the two photons exceeds the sum of the rest energy of the electron and positron.

The photograph shown in Fig. 5.4 presents a typical event corresponding to the diagram in Fig. 5.3. It is obtained using an instrument called the *bubble chamber*, which can be considered a direct successor of the cloud chamber. The operation of these two detectors is very similar, but the quality of images produced by the bubble chamber is much better. Instead of supersaturated vapor, the bubble chamber is filled with *superheated liquid*. The liquid in this state is almost ready to become a vapor, and the conversion is easily triggered by the passage of charged particles. Like in the case of the cloud chamber, small bubbles of vapor appear only along the trajectories of particles. The chains of these bubbles form clearly visible tracks, which are photographed and analyzed by scientists.

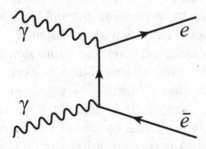

Fig. 5.3   Diagram of the creation of an electron and a positron from two photons colliding.

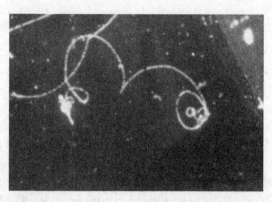

Fig. 5.4   Photograph of the creation of electron and positron by two (invisible) photons. Fragment of an image produced by a bubble chamber at Fermilab (USA).

The photograph in Fig. 5.4 shows the trajectories of two particles originating inside the volume of the bubble chamber. The particles are created at a single point and initially move in the same direction. A magnet used in conjunction with the bubble chamber deviates the particles to opposite sides. This means that the particles have opposite charges. Also, the particles can be unambiguously identified as an electron and positron using their path lengths. All these conclusions are quite straightforward, but the interpretation of the event presented as the process described by the diagram in Fig. 5.3 needs some explanation.

First, the diagram in Fig. 5.3 contains two vertices, while the actual event happens in a single point. As we already mentioned, this is a common feature of all processes in the microworld. Second, the photons do not leave traces in the bubble chamber since they are neutral particles. That is why the collision of two photons is not visible, and the photograph gives the impression that the electron and positron appear from nothing. Third, the two colliding photons are quite different. One of them is very energetic. It comes from outside the bubble chamber. The origin of the second photon is much more interesting. It is emitted by one of the charged nuclei of the chamber's material. Any charged object can do this. In general, the encounter of two photons is a rare event, but it happens easily when a photon passes in the vicinity of a nucleus. The only caveat is that the energy of the photon emitted by the nucleus is extremely small. Therefore, the first photon must bring all the energy required to produce the electron and positron. A fraction of this energy, which exceeds the rest energy of the two created particles, is converted into their kinetic energy. This is why the newborn particles begin their life by moving with considerable velocity, and the direction of their movement initially coincides with the direction of the first photon. But as soon as the electron and positron begin to exist, the magnetic force deviates their trajectories to opposite sides.

Thus, the event shown in Fig. 5.4 indeed displays the creation of an electron and positron by two photons. Most probably, this photograph does not give rise to any emotion, but in fact, it depicts an incredible phenomenon. It illustrates the conversion of radiation into matter. When a magician leisurely walking across a stage extracts a bunch of roses from thin air, we admire his prowess, but still, we are sure that he must have hidden the roses somewhere up his sleeve. In the event displayed in Fig. 5.4, something different happens. Shapeless and ethereal radiation transforms into a particle of matter. The same particles form everything around us. What is even more wonderful and amazing, is that the conversion is accompanied by the creation of an antiparticle. And all this is real, without any tricks. Undoubtedly, this is one of those ordinary miracles that could not even be dreamt of in philosophy just a few hundred years ago.

The electron and positron in Fig. 5.3 can be replaced by any other charged particle and antiparticle, so that all leptons and quarks, except neutrinos, can be produced in this way. On the other hand, the production of matter particles is not limited by the interaction of two photons. Many other colliding particles can do this. But all such processes must satisfy one strict condition — the conservation of energy. In other words, the energy of the initial colliding particles must be equal to or exceed the rest energy, which is proportional to the sum of masses of the created objects. Thus, the production of more massive particles. requires more energy.

As long as the energy condition is satisfied, all particles, including those not normally present in nature, can be created. The possibility of creating new particles is, of course, very exciting. When physicists realized this about sixty years ago, they started to explore the collisions of energetic particles. This kind of research was one of the primary goals of building accelerators. Very quickly, the construction of more and more powerful accelerators turned into a race, in which different institutions and even countries competed. The trophy in this race was obvious: being able to produce accelerated particles of higher energies opened more opportunities to discover unknown objects.[3] This race led to the invention of a new type of accelerator called the *collider*. In this facility, instead of directing the accelerated particles onto a static target, two beams of energetic particles move towards each other. As in an ordinary accelerator, this motion can be arranged inside a circular or linear pipe. The particles in each beam are accelerated to a very high energy. When the required energy is achieved, the beams collide in a specific location where an experimental detector is placed. Because of the head-on collision, all kinetic energy stored in the colliding particles can be converted into the rest energy of the created objects. It turns out that this conversion in head-on collisions is much more efficient than that in the collisions with a static target. That is why colliders have become the preferred instruments of particle physicists.

Experiments at colliders allowed scientists to find many new particles. For example, let us consider the discovery of the tauon. The corresponding experiment was carried out at the collider SPEAR, which accelerated and collided electrons and positrons. The discovery was made by a collaboration of scientists under the leadership of the American physicist Martin Lewis Perl. The diagram corresponding to the creation of a tauon is shown in Fig. 5.5. In this diagram, the initial state contains the colliding electron and positron. The tauon (τ) and its antiparticle ($\bar{\tau}$) appear in the final state. The collision can be treated as the annihilation of electron and positron because both

---

[3]In this connection, there is a good old joke. Two physicists walked in a forest and suddenly encountered a bear. They started to run away, the bear followed them. One physicist cried: "Wait, this is stupid. We cannot run faster than the bear." And the second answered: "I don't need to run faster than the bear. I need to run faster than you."

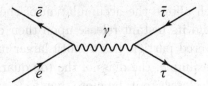

Fig. 5.5   Diagram of production of the tauon and anti-tauon pair in annihilation of electron and positron.

these particles disappear. Thus, annihilation can result in the production of different final states, not just of two photons as discussed before. Essentially, annihilation can produce any pair of particle and antiparticle, provided the initial energy of the electron and positron is larger than the rest energy of the created objects. In the language of diagrams, the symbol of tauon in Fig. 5.5 can be replaced by the symbol of any other fermion, and the process corresponding to the new diagram must also exist.

In the SPEAR collisions, the initial energy was sufficient to create several types of fermions that were known at that time, such as muons or strange quarks. Although the production of known leptons and quarks can be an exciting subject of study, the physicists were interested in the discovery of new particles. Here, the production of known fermions represented a considerable obstacle.

The problem is that it is impossible to predict what will materialize in each particular event of the collision. Therefore, all events with both known and unknown particles should be recorded. After that, the interesting events, which may contain new particles, should somehow be singled out from the rest of the sample. Usually, the interesting events are rare, while dull events involving well-known particles constitute the bulk of the selected sample. Thus, the work of experimenters can be described as the perennial task of separating good seeds from weeds. To a large extent, the result of any experiment depends entirely on the skill and ability of scientists to do this separation. Perl, with colleagues, succeeded in dealing with this problem and consequently discovered the tauon in 1974. This achievement brought Perl the Nobel Prize in 1995.

The discovery of the tauon nicely illustrates the modern technique of searching for new particles. The main instrument in this research is a collider, which allows the accumulation of energy in the accelerated particles and its instant release upon their collision. Usually, each newly discovered particle has a much larger mass than all preceding ones. For example, the mass of the top quark exceeds that of the bottom quark, closest to it, by more than forty times. Thus, more energetic collisions are needed to produce as yet unknown particles. That is why each newly built collider is much larger than its predecessors. The latest such facility, the Large Hadron Collider (LHC) built in CERN, represents a ring of more than eight kilometers in diameter tunneled underground at a depth of 175 meters.

The LHC accelerates and collides protons. By the end of the acceleration cycle, the accumulated energy of the protons exceeds their rest energy by several thousand times. Collisions with such huge energy have already yielded exciting results. In 2013, the ATLAS and CMS experiments at LHC reported the observation of a new particle called the *Higgs boson* which was predicted by theorists more than fifty years ago and had been searched for relentlessly since then. This discovery is just the beginning because the LHC is expected to operate for at least the next fifteen years. The next goal of the experiments at this collider is to identify the particles forming dark matter.

The discovery of new particles is by no means an end in itself. Particle physics has a much more ambitious goal. In essence, the collider experiments strive to understand the processes which formed the universe and determined its evolution. In the beginning, the universe was filled with a very energetic radiation, which created the particles and antiparticles through interactions similar to that shown in Fig. 5.3. These processes can now be modeled and studied in collider experiments, and this capability is very exciting and promising.

Figure 5.6 shows one of the collisions at the LHC registered by the experiment ATLAS. In this figure, the direction of the initial protons is perpendicular to the plane of the picture, and their collision occurs right in the center. The collision generates a large number of particles and antiparticles moving in all directions. Their tracks, which are registered by the ATLAS detector, are seen as curved lines. Events like this are like photographs of the universe taken in the first instants

Fig. 5.6　One of the collisions registered by the ATLAS experiment at the LHC (CERN).

of its existence. In effect, they display the great mystery of the birth of matter. Something similar happened billions of years ago when the collisions of the quanta of initial energetic radiation started to produce particles and antiparticles. That is why the colliders can rightfully be called the tunnels to the beginning of time.

The pictures taken at colliders will never be complete because the energy of collisions achievable by all current and future facilities will always remain lower than the energy of radiation that filled the universe immediately after the Big Bang. Thus, particle physics experiments can study the universe only from the moment when the universe cooled, with the energy of initial radiation decreased to a level comparable with that of the colliders. In any case, physicists aim to get as close to the instant of the Big Bang as possible. Therefore, increasing the collision energy is essential for their investigations.

Even though experiments at colliders have energy limitations, they reveal a lot about the creation of matter in the universe. One of the main results of this exploration is the conclusion that radiation can indeed be converted into matter. Scientists can reproduce such processes and study them. Thus, the origin of matter becomes less mysterious. But simultaneously with the clarification of one problem, research at colliders reveals a new one. In all current experiments, each materialized particle is accompanied by the corresponding antiparticle. This statement is confirmed by the diagrams in Figs. 5.3 and 5.5. Such behavior is also observed in all other processes studied so far. And so, it is natural to ask: where is all that antimatter which assisted the birth of matter? The answer to this question is yet unknown, and searching for it will be the subject of the remaining part of this book.

In science, it is always thus. After solving one problem, another puzzle, which follows from the obtained answer, immediately emerges. This chain of questions and answers never ends. But this does not mean that science is like the labor of Sisyphus.[4] After answering each new question, a net remainder is always left. There is a gain of knowledge, a better understanding of nature. This is the main scientific outcome, which allows humankind to develop and move forward.

. [4]In Greek mythology, the gods punished Sisyphus and forced him to roll a heavy boulder to the top of a mountain. But when the goal was achieved, the stone rolled back to the bottom and Sisyphus started his useless and unending labor again.

Chapter 6

Mystery of Disappeared Antimatter

6.1 Rule of Pairs

Particle physics has existed as a branch of science for a little over one hundred years. During this relatively short period, it has established many important laws that govern all processes occurring in nature. One such result is the *rule of pairs*. It states that each particle of matter is born and destroyed in a pair that includes its corresponding antiparticle. This rule is not relevant to the bosons transmitting interactions, such as photons, which can be emitted or absorbed in any quantity. However, leptons and quarks, which form material objects, strictly obey it.

It is fair to say that the rule of pairs is an *empirical rule*. That is, it is based only on experimental observations. Thus, its validity in all processes detected so far does not exclude its violation in some hypothetical interactions, which might be discovered in the future. Certainly, nature is full of surprises, which nobody can foresee in advance. Still, currently, not a single experimental result indicates that the rule of pairs could be wrong. That is why it should be considered as correct until the opposite is proven.

The rule of pairs works well in all electromagnetic processes considered so far. Its validity is also verified in other types of interactions. The most spectacular confirmation of the rule of pairs is demonstrated in the case of *weak interaction*, which was discovered at the beginning of the twentieth century. At that time, it was found that some nuclei can spontaneously emit an electron, simultaneously turning into nuclei of another element. Such a transformation was called β decay. A nucleus does not contain electrons; it consists of only protons and neutrons. Therefore, the electron must be produced by some interaction of the particles inside the nucleus. This interaction did not show up in any other phenomenon outside the nucleus; maybe because of this, it was called weak. β decay was the first manifestation of weak interaction. At the same time, it was the first process identified, in which a particle of matter was created, although scientists initially did not realize this.

According to the rule of pairs, the birth of an electron should be accompanied by the appearance of an antiparticle. But nothing was detected in β decay except the emitted electron. Nonetheless, at the beginning of the twentieth century, nobody worried about this, because neither antiparticles nor the rule of pairs was known at that time. More serious was the apparent violation of the law of energy conservation. The initial and final nuclei had different masses. Consequently, their rest energies were different. This difference was expected to be carried away by the emitted electron. However, the measured energy of the electron was much smaller than the value predicted by the law of energy conservation. The impression was that energy disappeared in β decay. Because of this missing energy, physicists became extremely agitated. As was stated by one scientist, "It's well better not to think about this at all, like new taxes".[1] It got to the point where Niels Bohr, the prominent theorist in nuclear physics, suggested that energy might not be conserved in nuclear processes. But another leading theorist Wolfgang Pauli put forward an opposite hypothesis. He supposed that the missing energy was taken away by a special neutral particle appearing together with the electron. For some reason, this particle escaped detection by scientific instruments. This could create the perception of disappearing energy, while in fact, energy continued to be perfectly conserved in β decay like everywhere else.

It should be noted that Pauli himself called his idea "a desperate remedy to save ... the law of conservation of energy." The violation

---

[1] This phrase is attributed to the Nobel laureate Peter Debye, who said this in a conversation with Wolfgang Pauli.

of this law seemed to him a much greater evil than the existence of a particle that could not be detected. In any case, this idea of Pauli's was accepted by the scientific community. The particle was named the *neutrino* and theorists started to explore its properties. Experimental physicists kept pace and began to develop methods to catch the neutrino. Finally, after several dozens of years of continuous research, the particle was experimentally detected. An important outcome of this quest has been the impressive confirmation of the law of energy conservation and its establishment as an inviolable rule governing nature.

Notwithstanding this tremendous success, there was a small inaccuracy in the hypothesis of Pauli, which neither he nor anybody else appreciated for a long time. The particle appearing in β decay is not a neutrino, but an antineutrino. More precisely, this particle is an *electron antineutrino* which is always created together with an electron. A neutrino, like any other lepton or quark, has its antiparticle. The charge of a neutrino is zero. Therefore the neutrino is very difficult to distinguish from the antineutrino. Nevertheless, it is possible to prove that the invisible object, which was emitted in β decay together with the electron, is the antiparticle.

The appearance of the antineutrino in β decay was first established in 1956. The experiment was carried out at the nuclear reactor of the Savannah River Plant in the USA. This plant produced material for nuclear bombs and was therefore a highly secured facility. However, a group of scientists directed by Clyde L. Cowan and Frederic Reines was allowed to conduct non-military research there. β decay is one of the main processes taking place in any nuclear reactor. Also, the amount of radioactive material utilized in the reactor is huge. Thus, the number of β decays, which occurs there every second, is enormous, and the flow of the emerging antineutrino is strong. That is why the scientists decided to set up a detector of antineutrinos near the nuclear reactor. The antineutrino was expected to interact with the detector material and thus reveal itself.

Both the neutrino and antineutrino very rarely interact with other particles. Even if all the space between the Sun and the Earth were filled with iron, most of the neutrinos being created in the Sun would travel the whole distance to Earth without hitting a single atom. Starting from the first mention of the neutrino by Pauli and for a very long time since then, the neutrino was regarded as an entirely undetectable particle. Cowan and Reines themselves called their experiment "Poltergeist", so elusive the neutrino seemed to be. Its detection was possible only because of an essential property of weak interaction, which controls all processes involving the neutrino. This interaction changes the flavors (that is, the types) of the fundamental fermions participating in it. For example, a neutrino colliding with a nucleus turns into an electron. An antineutrino, in its turn, becomes a positron. Thus, a neutrino and antineutrino can be identified by detecting an electron or positron, respectively, which emerge from their interaction. At the same time, the difference between the electron and positron allows scientists to distinguish the original neutrino from the antineutrino.

Because of the change of flavors occurring in weak interaction, Cowan and Reines hoped to register the following chain of processes. The antineutrino, which was created in the nuclear reactor, should interact with a proton contained in the material of the detector, and transform into a positron. The proton after this interaction should become a neutron. The subsequent annihilation of the positron should produce two photons.

Thus, the interaction of an antineutrino was expected to produce two photons and a neutron. The photons could be easily detected. The appearance of the neutron was also possible to identify. For this, Cowan and Reines added a special chemical substance into the composition of the material of their detector. The nuclei of this substance absorbed neutrons and emitted several additional photons. This emission was slightly delayed with respect to the first two photons from the annihilation. As a result, a typical interaction of the antineutrino produced two bursts of photons separated by a certain time interval. None of the *background processes*[2] could imitate such a signal.

---

[2]The background process is a process that is not related to the phenomenon of interest, and which impedes its detection.

The events corresponding to the expected antineutrino interaction were indeed detected. Moreover, they were registered only when the reactor operated. This observation proved the existence of the antineutrino. Reines was awarded the Nobel prize in 1995 for this discovery, but Cowan, unfortunately, died before this triumph.

The discovery of Cowan and Reines not only confirmed the existence of the antineutrino but also nicely demonstrated the validity of the rule of pairs. The production of the electron in β decay is indeed accompanied by the simultaneous creation of an antiparticle. If a neutrino (that is, a particle) were created there, it would produce an electron in the interaction with the detector material. The electron would not cause annihilation. Consequently, two bursts of photons separated by a time interval would not be registered.

This example of weak interaction, in which an electron appears together with an antineutrino, shows that the rule of pairs allows for the creation of a particle and antiparticle with different flavors. What is important is that the same number of particles and antiparticles emerges. After an adjustment for the change of flavors, all known processes obey the rule of pairs. The number of events registered and studied in experiments at colliders amounts to billions, and particles and antiparticles are produced strictly in pairs in all of them. The theoretical description of all known processes also respects the rule of pairs. For example, the laws of electromagnetic interaction do not allow the creation of an electron without a positron.

The rule of pairs has a significant consequence related to the creation of matter during the evolution of the universe. If this rule is right, then the generation of each new particle should be complemented by the birth of an antiparticle. Therefore, equal amounts of matter and antimatter should be present in the universe. But is this indeed the case? This question can be answered only by attempting to find antimatter somewhere, here on Earth or in a distant galaxy. These efforts and their results are considered in the next section.

6.2   Searching for Antimatter

According to the rule of pairs, antimatter should be abundant in the universe. Our galaxy, the Milky Way, contains more than three hundred billion stars. The total number of similar or even larger star formations in the observable universe is about two hundred billion. If matter and antimatter had been produced in equal proportions, then half of this immense number of stars and galaxies should be made of antimatter. However, it is impossible to tell whether a star is made of antimatter just by looking at it from a distance. Antiparticles are expected to obey all the laws that particles do, and therefore all nuclear reactions that occur deep within regular stars must be reproduced by antinuclei igniting antistars. Thus, to an external observer, an antistar should appear identical to an ordinary star, and the presence of antimatter within any celestial object can only be deduced indirectly.

Here on Earth, it is a different case. The question of the existence of antimatter on our planet is straightforward to answer, and the answer is perfectly unambiguous. Earth must be composed of only ordinary matter. If half the Earth were ever made of antimatter, we would all be photons by now. Even if the space dust, from which the Earth formed several billion years ago, had contained any fraction of antimatter, that antimatter would have been annihilated a long time ago. Cosmic rays can bring antiparticles to the Earth from the depths of space, but they are destroyed almost instantly in the upper layers of the atmosphere, which is full of ordinary particles. Another source of antiparticles is colliders, where billions of antimatter particles are produced during scientific experiments. However, after decades of operation, all of the world's experiments put together have generated less than one-millionth of a gram of antimatter. Even this tiny amount disappears sooner rather than later because of annihilation. There are no other significant sources of antiparticles on our planet.

The next question is whether the Sun contains any antimatter. The answer is also evident: no. We know this because of the continuous flow of charged particles emitted by our star together with its light. This flow is called the *solar wind*. One of the sources of this wind is colossal ejections of matter during solar flares. These ejections can easily throw off billions of tons of solar material into space in a single burst. But even without the flares the solar wind never stops. On average, it carries away billions of kilograms of solar matter per second. The particles of the solar wind move at high speeds sufficient to escape the Sun's gravity. They travel in all directions, and some of them reach the Earth. Here, they are deflected by the magnetic field of our planet towards the north and south poles. At the poles, they collide with molecules in the air and produce beautiful and breath-taking polar lights called *Aurora Borealis* and *Aurora Australis*, which can be best observed during winter nights.

If the Sun were made of antimatter, then the solar wind would contain antiparticles, such as antiprotons and positrons. For the Earth, being a neighbor of an antistar would have catastrophic consequences. The mass of our planet would continually decrease as the particles on Earth would annihilate with the antiparticles of the solar wind. Furthermore, instead of the colorful polar lights, the surface of the Earth would be showered with deadly rays of photons produced during annihilation.

The annihilation of any antiparticle results in the creation of several photons. Two photons are generated in the annihilation of a positron while the number of photons produced in the annihilation of an antiproton is much larger. The danger caused by photons depends on their energy. Low-energetic visible light is completely harmless for humans and is even essential for the synthesis of vital vitamin D required for the development of our bodies. Ultraviolet radiation has a higher energy. It can cause sunburns and even cancer in the case of prolonged exposure. The energy of photons produced in annihilation is hundreds of times larger than that of ultraviolet rays. Photons emitted in the annihilation of antiprotons are particularly energetic and are therefore especially dangerous. This type of radiation does not just burn the skin — it kills. The photons penetrate deep inside the body, destroying cells and DNA. Therefore, if the solar wind contained antiparticles, the photons produced by annihilations would quickly and inevitably kill any organism on Earth. Moreover, it would have been impossible for life to begin in such hostile conditions. Thus, the mere fact that we exist provides strong evidence that the Sun contains only ordinary matter.

The presence of the solar wind also proves that the solar system is made up only of ordinary matter. The solar wind disappears only at about 17.5 billion kilometers from the Sun, as detected by the space probe Voyager 1 in 2010. This distance is well beyond the orbit of Pluto. Up to these limits, the solar wind bombards every object in the solar system. Nothing can hide from it. It is because of the solar wind that the tails of comets point away from the Sun regardless of the direction of the comet's motion. If there were any objects within the solar system made of antimatter, its antiparticles would annihilate with particles of the solar wind, and would emit numerous energetic photons. Such photons are seldom produced by any other space objects, and they would, therefore, be easily detected by modern astronomical instruments. But nothing like this has ever been observed. The flow of energetic photons has never exceeded the value expected from conventional sources. Thus, we can confidently say that there is no antimatter wherever there is the solar wind, that is, in the entire solar system.

Any assertion about the existence of antimatter beyond the bounds of the solar system is harder to make. However, it can still be done thanks to "special messengers" from faraway stars, the *cosmic rays*, introduced earlier. The origins of high-energy particles composing the cosmic rays vary. Some of them arise from *supernovae* explosions. Such explosions happen at the end of the life of many stars. Supernovae emit so much light that they can briefly outshine an entire galaxy. During a supernova explosion, almost all the matter contained in the star is ejected into the surrounding space with speed close to that of light. Some of this matter reaches the Earth in the form of cosmic rays. Another source of cosmic rays is thought to be the supermassive black holes located at the center of many galaxies. Such black holes act as giant centrifuges which are capable of accelerating charged particles to extremely high energies and shooting them in all directions. Still, to arrive at our planet, the envoys of remote galaxies need to travel millions of years and cover billions of kilometers. How many events, that have happened in the universe, they have witnessed! How much they could tell us if only they could speak! Of course, particles cannot speak. Instead, they leave traces in scientific devices, and those who can decode these fleeting tracks, these silent messages from distant worlds, can uncover the greatest secrets of the universe.

In the search for antimatter, the most important and relevant property of cosmic rays is that they originate from beyond the solar system. According to numerical estimates, supernovae happen about three times every century in our galaxy. The Milky Way is over 13 billion years old. Thus, the number of supernovae that have occurred during this period is so large that the flow of cosmic rays produced by them never decreases and continuously pounds the Earth. If some of the stars in the galaxy were made of antimatter, then some of them would also explode as supernovae, because all the physical processes that govern the life of antistars are almost the same as those occurring in ordinary stars. These explosions would propel the material of the antistar into space and, as a result, some of its antiparticles would reach the Earth. Therefore, we could detect the existence of antistars by finding antiparticles in cosmic rays.

However, not every antiparticle detected in cosmic rays provides unambiguous evidence for antistars. Positrons, for example, are frequently created when ordinary particles in cosmic rays collide with interstellar material. The production of an electron-position pair does require a sufficiently high energy. But this condition is easily satisfied since particles in cosmic rays are usually highly energetic. Therefore, while cosmic rays indeed contain a significant fraction of positrons, this does not prove anything. The probability of an antiproton being produced in collisions with the interstellar material is also rather large for the same reason, and so identifying antiprotons in cosmic rays does not evidence anything either.

The most convincing finding would be detecting antihelium. The antihelium nucleon consists of two antiprotons and two antineutrons. Hence, to obtain the antihelium nucleon, the four antiparticles would need to be created simultaneously in a single collision and then bound together immediately after their production. Such a process almost never happens, so that antihelium is rarely generated in particle collisions. If antihelium exists in interstellar space, its source can only be the supernova explosion of an antistar.

Helium is one of the primary materials stars are made of. In fact, helium was first detected not on Earth, but in the Sun, and it is named after Helios, the Greek God of the Sun. Its abundance in stars is about 25% because this element is the main product of the nuclear fusion taking place inside them. A supernova releases all the helium contained within it. Similarly, an anti-supernova should free all of its antihelium. This released antihelium would form part of the cosmic rays. This is why even a small fraction of antihelium in cosmic rays would be a clear indication of the existence of antistars.

Studying cosmic rays is not a simple task because the atmosphere of the Earth does not allow them to reach the surface. All of the particles in cosmic rays interact with molecules in the air and quickly disappear, producing showers of secondary radiation. Even this secondary radiation gradually diminishes as it approaches the surface. Consequently, almost all information carried by cosmic rays is lost at the Earth's surface. The atmospheric shield is beneficial for life on our planet because it protects us from the destructive action of cosmic rays. But it also makes studying cosmic rays in a surface-based laboratory impossible. In fact, cosmic rays were only discovered when a device measuring the level of natural radiation was placed on a balloon and launched up to an altitude of over 5000 meters. This experiment was carried out by the Austrian-American physicist Victor Hess. He observed that radiation increased considerably at high altitudes and concluded that it must penetrate the atmosphere from above. His discovery paved the way for a comprehensive study of cosmic rays using scientific instruments sent into the upper layers of the atmosphere and even into outer space.

The best measurement completed so far of the fraction of antihelium in cosmic rays is due to the BESS experiment performed by a collaboration of Japanese and American scientists. They used a special balloon capable of flying at an altitude of about forty kilometers. For comparison, this is much higher than the height of Mount Everest (8848 m) or the altitude at which passenger aircraft fly (about 10 km). The BESS experiment aimed to detect and identify the particles that comprise cosmic rays. After several launches of their balloon, the scientists registered more than 40 million nuclei of helium in the cosmic rays, but not a single nucleon of antihelium. Finding helium is not surprising — it aligns with the expectation that supernovae eject this element into cosmic rays. Similarly, not finding antihelium means just one thing: that antistars are very rare in our galaxy, or perhaps even absent.

As is always the case, a negative result is still a result. The absence of antihelium compared to the number of helium nuclei detected allows for the estimation of the maximum possible fraction of antistars in the galaxy. Detailed calculations based on the BESS results suggest that there can be no more than one antistar per ten million stars. Nonetheless, this does not mean that there is an absolute absence of antimatter in our galaxy. The Milky Way contains about 300 billion stars. Therefore, there might still be a few thousand antistars. Detecting just one antistar would be a significant discovery. It would tell us a lot about the composition of the universe and its evolution. Hence, the search for antihelium in cosmic rays continues.

Besides BESS, several other experiments are seeking antihelium in cosmic rays. One of these experiments is very special. It is performed in outer space on board the International Space Station (ISS). One of the last missions of the space shuttle launched in 2011 carried a sophisticated instrument to the ISS. This device can determine types of particles in cosmic rays. It has been successfully installed on the external surface of the space station. Using this instrument, the AMS experiment, which studies cosmic rays, is currently being conducted. The sensitivity of AMS is much greater than that of any other experiment carried out within the Earth's atmosphere.

Recently, the first AMS results on the search for antihelium have been released. In five years, AMS has collected 3.7 billion helium nuclei, which is about a hundred times more than the amount accumulated by BESS. What is more exciting, the experiment has also recorded several antihelium nuclei. No conclusion has been derived from this result yet. The scientists involved in such research have stated that they need some time for a detailed understanding of the instrument to reject any possibility of background contribution.

Depending on the outcome of this thorough verification, AMS will improve the upper limit of the maximum possible number of antistars in the galaxy by a factor of hundred, or it will prove that antistars exist. Of course, the second possibility would be much more exciting.

In any case, it is now obvious that antimatter is very rare in the Milky Way. Its fraction is much less than the 50% expected from the rule of pairs. However, the absence of antimatter in our galaxy does not mean that there is no antimatter elsewhere in the universe. Therefore, scientists have also searched for antimatter beyond the Milky Way.

After galaxies, the next-largest structures in the universe are *galaxy clusters*. These are giant aggregations of individual galaxies that can include up to several thousand galaxies bound together in a relatively small volume by the force of gravity. Galaxy clusters contain an enormous amount of matter. However, it turns out that atomic matter (the only known form of matter so far) constitutes just ten percent of the total mass of galaxy clusters. The remaining mass is in the form of a mysterious dark matter about which very little is known. Conversely, the atomic matter in galaxy clusters is studied quite well. The majority of it is contained not in the galaxies, as might be naively expected, but in the so-called *intracluster medium*. This refers to a plasma located in the center of the cluster which is heated to a temperature of up to hundreds of millions of degrees by the pressure of the gravitational force. Due to this extremely high temperature, the plasma emits X-rays, which can be detected by special telescopes orbiting the Earth on board satellites.

Naturally, after galaxies, galaxy clusters are the next place to search for antimatter. If antimatter were present in the intracluster medium, it would annihilate with ordinary particles and release photons. These photons act as "signatures" of antimatter. Their energy is much higher than that of the X-ray photons, so they cannot be mistaken for something else. Since the X-ray radiation of the intracluster medium reaches the Earth, the photons from the antimatter annihilation should be detectable here too. Space telescopes would register them along with X-ray radiation. Over the last several years, 55 galaxy clusters emitting X-rays have been studied, and scientists still have not observed any excess in energetic photons. Hence, we can conclude that the fraction of antimatter in the intracluster medium is tiny. According to numerical estimates, there can be no more than one antiatom per million ordinary atoms.

To summarize the presented evidence, antimatter is almost excluded at the scale of the Earth, the solar system, the Milky Way and in galaxy clusters. At all these levels, only matter is observed while the maximum allowed fraction of antimatter is negligibly small. It could be supposed that some galaxy clusters are totally made of antimatter. But a detailed theoretical study excludes this possibility too. Domains of matter and antimatter cannot be totally isolated from each others, and therefore their boundaries must touch. Annihilation would occur at these boundaries, and detectable photons would be produced. However, scientists have never observed such photons. We are therefore forced to conclude that if a region made entirely of antimatter exists anywhere, it must be located beyond the so-called *observable universe*.

The observable universe is the part of space containing stars and galaxies from which a signal, such as light, can, in principle, reach the Earth during a time interval within the lifetime of the universe. The distance to the edge of the observable universe is estimated to be about 46 billion light years. We cannot know what happens beyond this distance because the signals emitted there have not had enough time to reach us even if they had been sent in the first instants after the Big Bang. This is a fundamental barrier which no telescope, no matter how big or sophisticated, can penetrate. If antimatter is hidden beyond the bounds of the observable universe, we will never be able to discover it. However, the impossibility of verifying this hypothesis makes it virtually meaningless; science prefers to deal with statements which can be tested at least in principle. Furthermore, even if the hypothesis were correct, we would still have to understand what had caused all antimatter to be collected exclusively in a place that we cannot see from Earth. This is something that is just as difficult to explain as the absence of antimatter in our observable universe. In fact, it just replaces one mystery with another.

6.3   The Problem Well Put

Let us summarize the results obtained so far. The continuous and numerous investigations of extremely small and incredibly large objects coexisting in the universe lead to the following two conclusions. **First**, based on the study of stars and galaxies, we can confidently claim that antimatter is very rare or even utterly absent in the observable universe. **Second**, the exploration of the microworld of particles provides an equally compelling proof that matter and antimatter are created and destroyed in equal proportions. These two results, considered separately, are supported by a solid evidential basis, and therefore seem undeniably correct. However, taken together, they apparently contradict each other.

There are many different attempts to resolve this paradox. One possibility is to invalidate either the first or the second statement and thereby eliminate the contradiction between them. For example, some theories suppose that antimatter does not form antistars and antigalaxies. That is why astronomical instruments do not detect the presence of antimatter, although the abundance of matter and antimatter in the universe is in fact equivalent. Other theories suggest that the Big Bang has created only particles of matter due to some yet unknown process. However, all such theories require many additional assumptions. For example, the theory explaining the absence of antistars needs to introduce some specific property, which prevents antimatter from forming antistars and does not affect ordinary matter. This is quite unnatural because all properties of matter and antimatter studied so far are essentially the same. In any case, none of such assumptions are confirmed experimentally.

There is also another possibility. Suppose that initially, the number of created particles and antiparticles had indeed been the same, but later, during the subsequent evolution of the universe, all antiparticles vanished, for some reason. If this disappearance happened before the formation of antistars, it explains why we do not detect them now. Of course, to implement this possibility, we need to answer one more question: **why has antimatter disappeared**? Solving this new problem seems very difficult. But difficult does not mean impossible. In any case, particle physics attempts to do this. More-over, explaining the disappearance of antimatter is listed as one of the main issues that particle physics needs to address.

The next parts of this book describe all that has been achieved in this direction. Summarizing these achievements, we can positively state that much has already been done. Many of the results obtained have revealed totally unexpected properties of matter, which have completely changed our view of nature. One of the main results of this quest is the conclusion that the smallest particles are indeed the key to understanding the immense universe. However, the full explanation of the origin of the absence of antimatter is not yet known. This mystery still needs to be solved. But we should not despair. What is important is that we can ask the question and we understand its meaning. As was stated by the American philosopher and psychologist John Dewey: "the problem well put is half solved." And so, it now remains just to uncover the second half.

PART 2
Broken Symmetry

"Curiouser and curiouser!"
Lewis Carroll, *Alice's Adventures in Wonderland*

Chapter 7

Baryogenesis

The beginning of the universe was humble compared to its current splendor and greatness. Initially, it had nothing except energy compressed in an incredibly small volume. It took a while for matter to appear.

The processes preceding the creation of matter remain hidden from us. An abyss of time, which separates us from those events, seems so immense and unbridgeable, that we can only guess at possible mechanisms of the universe's evolution during the first instants of its existence. However, several theories that describe the birth of the universe already exist, and numerous experimental observations have confirmed some of them. According to one such theory widely supported by scientists, the initial expansion of the universe was extremely swift. More precisely, the space of the universe rapidly increased. The reason for this growth was an enormous density of energy that filled the universe. Figuratively speaking, energy inflated the universe from inside out and steeply enlarged its dimensions. Because of this analogy, the corresponding theory is called the *inflationary model*.

As the universe swelled, the energy density reduced, and the expansion of the universe slowed. At last, the environment became favorable for the appearance of the first particles of matter. The conversion of energy into matter and the subsequent transformations of particles are currently well-known and are modelled in collider experiments. We considered some examples of these transformations in Chapter 5. Therefore, starting from the emergence of matter, the haze of uncertainty, which covered the evolution of the universe, began to dissipate, and the description of processes that happened then have become less ambiguous.

The original types of matter that arose from the sea of energy were leptons and quarks. Twelve flavors of these particles are known, and all of them were created equally well at the early stage.[1] Each generated particle was accompanied by its corresponding antiparticle. The density of particles and antiparticles in the young universe was so high, that they collided with each other almost immediately after being born, underwent annihilation and released energy spent on their production. In the next moment, this energy was again transformed into particles and antiparticles. The universe boiled, and matter with antimatter appeared and disappeared again and again.

The current structure of matter differs considerably from its initial form.

First, from twelve known flavors of fundamental fermions, only up and down quarks, electron and all three types of neutrino can exist indefinitely. All other quarks and leptons are short-living particles, which decay sometime after their production.

Second, all initial up and down quarks are now confined inside protons and neutrons. In their turn, protons and neutrons are united in atomic nuclei. Nuclei and electrons form atoms, the building blocks of all known substances.

And finally, **third**, the fraction of existing antiparticles is considerably smaller than the 50% expected from the rule of pairs. Certainly, no significant amount of antimatter is present in the observable universe. Because of this, matter does not annihilate and therefore remains in a stable condition.

The transformation of the initial mixture of particles and antiparticles into the current structure of matter is called *baryogenesis*. This term means literally "creation of baryons." The emphasis on *baryons*

---

[1]Besides quarks and leptons, the particles forming dark matter also had to appear at that time. But we know nothing about them, so any discussion about them would be premature.

is important, since baryons, or, more precisely, protons and neutrons, constitute the basis of all matter surrounding us. The theory of baryogenesis must explain all stages of the long path from the original soup of quarks, leptons, and their antiparticles to the present world where only protons, neutrons, and electrons exist while the presence of antibaryons and positrons is not noticeable.

Some stages of baryogenesis have already been thoroughly explored and have become mostly clear. The other phases of matter transformation still wait to be understood. The primary difficulty in composing a comprehensive picture of baryogenesis involves trying to explain the absence of antiparticles. Apparently, the action of some special forces of nature caused antimatter to vanish and matter to remain. These forces are unknown. However, finding them is crucial for comprehending the universe's evolution.

The disappearance of antimatter is only possible if properties of particles and their interactions obey several conditions. We will discuss these conditions later, but one of them can be stated already now. **The properties of particle and antiparticles must be different**. Only then can nature select a winner in the fight between matter and antimatter. This difference was out of the question for a long time. Starting from the first theoretical papers by Dirac, scientists expected the same behavior of particles and antiparticles in all interactions. Nonetheless, it turns out that the symmetry between particles and antiparticles is indeed broken. But to conclude this, we first need to consider all ingredients of baryogenesis. We do this in the following chapters. We also discuss many important properties of particles necessary for understanding why the differences between particles and antiparticles are possible and how they reveal themselves. We start the second part of our story by examining the interaction that forces all quarks of the universe to remain hidden inside baryons.

Chapter 8

From Quarks to Atoms

8.1 Strong Interaction

Quarks are charged particles and can, therefore, interact electromagnetically. Besides, they can also participate in another type of interaction, which is called *strong*. These two types of interaction acting together determine the atomic structure of matter. Although some of their features are common, the strong and electromagnetic interactions are in fact very different.

Effects of the electromagnetic interaction are quite easy to observe and convenient to study. Even ancient Greeks were able to experiment with electricity when they rubbed a piece of amber and attracted feathers of birds to it.[1] By the end of the nineteenth century, the electromagnetic interaction was known so well, that Maxwell successfully derived his famous equations, which united electricity and magnetism. The strong interaction is stealthier. It is like a formidable beast, a lion or a tiger, safely locked behind bars. Or, it can be compared to an almighty genie kept inside a lamp by a magical spell. But contrary to all animals and genies, the strong interaction is never meant to break free beyond the limits of its prison. Manifestations of this interaction are present everywhere. It is the strong

---

[1] Amber in Greek is called "elektron." Thus, innocent amusements with amber carried out several thousand years ago gave the name to the phenomenon of electricity.

interaction that glues particles together inside nuclei and does not allow nuclei to break apart. But the direct participants of the strong interaction, the quarks, are hidden forever inside protons and neutrons, and the interaction itself never goes out beyond the edges of a nucleus. That is why nobody suspected the existence of the strong interaction until a nucleus was discovered.

Scientists realized the need for a special force acting between particles only after finding out that a nucleus contains many positively charged protons. For example, the nucleus of iron includes 26 protons, and heavier nuclei are comprised of even more of them. This observation contradicts a well-known law, discovered by the French scientist Charles-Augustin de Coulomb in the eighteenth century. Coulomb's law states that two objects with like charges repel each other. The smaller the distance between the objects, the stronger the force of repulsion. This law does not allow for any exception, and protons in nuclei must obey it also. The charge of each proton is very small, but the distance separating the protons in a nucleus is so tiny, that the force of electric repulsion would surely tear apart any nucleus and disperse the protons in all directions. Hence, according to Coulomb's law, no single object could exist. In reality, almost all nuclei are so stable that enormous efforts are needed to split them into parts.

To eliminate the evident discrepancy between the law of electricity and reality, physicists introduced an additional interaction. Its purpose was to hold together all neutrons and protons inside nuclei. The interaction was called "strong" because protons must attract each other with a force exceeding their electric repulsion. There are also several other conditions constraining the behavior of the strong interaction. It should not depend on the charge of particles because neutrons do not have charge, and yet they are held in nuclei together with protons. On the other hand, electrons must be insensitive to the strong interaction. Otherwise, instead of forming the shell of an atom, they would be pulled inside a nucleus and would remain there together with protons and neutrons. Finally, the strong interaction should be cut off abruptly outside a nucleus. Otherwise, the protons and neutrons from neighboring nuclei would attract each other and produce one big lump, so that atomic matter would not exist.

Because of these unusual features, the understanding of strong interaction was strenuous. For a long time, its study raised more questions than it gave answers. In the middle of the twentieth century, scientists already developed the comprehensive theory of electromagnetic interaction, the QED, which correctly explained all electromagnetic phenomena. The structure of this theory is rather simple, and its main features are the following. **First**, the electromagnetic interaction is conducted by bosons (photons). Each act of interaction consists of photon exchange between particles. **Second**, strict and well-defined laws determine this exchange. **Third**, particles participate in the electromagnetic interaction because they have a special property called electric charge. Consequently, uncharged particles don't interact electromagnetically. The success of the electromagnetic theory based on these principles was impressive. That is why there was a burning desire to build the theory of strong interaction in a similar manner.

In one episode of the adventures of the little girl, Alice, in the Wonderland described by Lewis Carroll, Alice fell into a pool of her own tears. Finding herself in the unusual situation, she tried to compare it to her previous experience. She decided that she had fallen into the sea and recalled that a sea meant a beach, children digging in the sand, lodging houses and a railway station. After that, the unknown situation became much clearer. Alice concluded that she could get out of the pool of tears by train. Likewise, physicists tried to explain the strong interaction by comparing it to what they knew well, the electromagnetic interaction. They thought that if they could establish a similarity between these two forces of nature, understanding the strong interaction would be simplified.

With this idea in mind, theorists attempted to build the theory of strong interaction in the following steps. First, fundamental bosons conducting the strong interaction should be identified. After that, laws that control the action of these bosons must be determined. Also, the theory should uncover a special property of protons and neutrons, something like a charge, which allows protons and neutrons to interact strongly. At the same time, the theory should explain why electrons don't have this property. Obviously, nothing from this long wish list was initially known.

In 1935, the Japanese theorist Hideki Yukawa put forward the first idea of bosons of the strong interaction. For this role, he proposed a special particle, which behaved similarly to photons in the electromagnetic interactions, although contrary to photons, Yukawa's boson had mass. Yukawa even predicted the value of this mass. Protons and neutrons could emit and absorb Yukawa's boson, while electrons were not sensitive to it. The reason for this split was not known; Yukawa just postulated this property.

The search for a particle corresponding to Yukawa's boson continued for more than twelve years with a break during World War II. Finally, a group of scientists from the University of Bristol in England under the leadership of Cecil Frank Powell discovered a likely particle. It was named a π *meson*, or just a *pion*. The pion possessed all required properties. This particle indeed belonged to the class of bosons contrary to all other particles (electron, proton, neutron, and muon) known at that time. It participated in the strong interaction. Also, its mass was consistent with the value predicted by Yukawa. Such a remarkable agreement between theory and experiment gave hope that the strong interaction would be fully understood soon. The enthusiasm of physicists was great. Everywhere, at scientific conferences and in laboratories, the unspeakable impression of a harmony in nature, in which all forces have a similar arrangement, was in the air. It was not surprising that Yukawa was awarded the Nobel Prize in 1949, and Powell received the same award one year later.

However, the theory, which employed pions as the fundamental bosons of the strong interaction, soon started to encounter significant problems. Numerous new results arriving from the first accelerators continually eroded its basis. In the sixties, scientists began to accelerate protons and collide them with targets containing protons and neutrons. These collisions were governed by the strong interaction, which thus revealed itself in all its strength. The collisions indeed produced many pions, which were thrown off from the collision point like sparks. This effect was expected since pions were thought to be the bosons of the strong interaction and could, therefore, appear whenever the strong interaction was involved. But in addition to pions, many other unknown particles emerged. All of them existed for a very brief time and quickly decayed. Their *lifetime* (that is, their period of existence) was much shorter than that of pions. Pions were also unstable, but their lifetime was long enough to leave tracks in scientific instruments. On the contrary, the new particles decayed almost instantly after their birth. In all other respects, the new particles were very similar to pions. They were also called *mesons* by analogy with pions to reflect this similarity. All new mesons belonged to the class of bosons like pions. Protons and neutrons strongly interacted equally well with both pions and other mesons. Therefore, all mesons could be treated as the fundamental bosons of the strong interaction. This situation was entirely different from the electromagnetic interaction, where only one type of boson, the photon, existed.

The list of baryons, which initially included only a proton and a neutron, also increased. Almost simultaneously with mesons, experimental physicists started to detect new particles that were rather like the proton and neutron, but had a larger mass. Because of this similarity, they were also called baryons. Each new baryon was a short-living particle and decayed to a proton or a neutron and several mesons. Otherwise, the properties of all baryons were quite similar.

All discovered mesons and baryons were found to be rightful participants of the strong interaction. Because of this feature, they were united into a single group and were called *hadrons*. Like many other terms in particle physics, the origin of this word is Greek. It is derived from the Greek word meaning "large", "heavy." This name opposes mesons and baryons to leptons, whose name means "small", "light." In fact, some hadrons are much lighter than the heaviest lepton, the tauon. However, the classification of particles into hadrons and leptons (see Fig. 8.1) depends not on their mass, but on their sensitivity to the strong interaction. Only hadrons participate in the strong interaction, while all leptons are neutral to it. Indeed, not only the electron but also the muon, the tauon, and all three neutrinos do not care about the strong interaction. The reason for this neutrality was obscured for a long time. In analogy with the electromagnetic interaction, scientists have supposed that all hadrons have some unique property, which allows them to interact strongly, while leptons don't have it. But the property itself remained unknown.

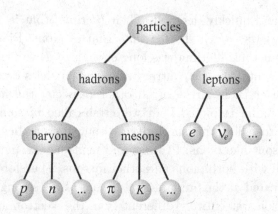

Fig. 8.1   Classification of particles.

With time, the number of discovered hadrons was snowballing, and it was becoming more and more difficult to understand what was what in the strong interaction. Yukawa's theory was initially neat and simple. In this theory, there were two baryons (a proton and a neutron), which exchanged bosons of just one type (a pion). But gradually this elegant theory turned into a mash where dozens of mesons and baryons were blended into an incomprehensible mixture and interacted with each other.

Nevertheless, a large volume of accumulated data on particle behavior in the strong interaction was of great importance. Each branch of science evolves through several stages. First, an extensive amount of information on the studied subject is collected. After that, the data are ordered and systematized. This systematization often results in an entirely new and much deeper understanding of the subject. This was the case in chemistry, where a large number of identified chemical elements led to the creation of the periodic table by the Russian scientist Dmitri Mendeleev. Similarly, in biology the English naturalist Charles Darwin used the information gathered on animals and plants to create his theory on the origin of species. The same scientific miracle happened in particle physics. In 1964, the American scientists Murray Gell-Mann and George Zweig proposed a *quark model*, which ordered all hadrons into a single coherent system.

8.2   Quark Model

Gell-Mann and Zweig suggested, independently of each other, that all hadrons were made of special particles called *quarks*.[2] Within this hypothesis, the observed diversity of hadrons was achieved by stating that quarks have different types, or *flavors*. Initially, just three flavors, which were named *up*, *down* and *strange*, were sufficient to explain the structure of all known hadrons.

The word "quark" had not existed in the everyday English language. Gell-Mann took it from the work *Finnegans Wake* by the Irish novelist and poet James Joyce. This work can be characterized as one big puzzle because it consists of a mixture of hardly understandable phrases. In particular, the work contains the nonsensical sentence "Three quarks for Muster Mark." The story is silent about who Muster Mark is. At least, Joyce never mentions him again. One version of the origin of this phrase claims that Joyce had once heard at a German market that a vendor advertised his produce by the slogan "Drei mark für muster quark." The motto can be translated from German as "Three marks for excellent curd cheese" (the mark was the German currency for a very long time). So, supposedly Joyce invented his famous phrase just by changing the order of words in the mundane sentence. Of course, there is no agreement on this subject. Several generations of linguists continue to cross swords trying to decipher the meaning of three quarks for Muster Mark. In any case, it is precisely because of this phrase that the same German word is used to name both the fundamental particles of matter and curd cheese. In *Finnegans Wake*, the word "quarks" was associated with the cry of gulls. Gell-Mann, in his turn, decided that this word fit well to the particles invented by him. Especially, he liked that the number of quarks in *Finnegans Wake* was equal to three, like in his theory. Later, the number of quark flavors was extended to six, but the number three remained tightly related to quarks, although not in the same way as it had been expected by Gell-Mann.

---

[2]Zweig named the new objects "aces", but this term did not survive.

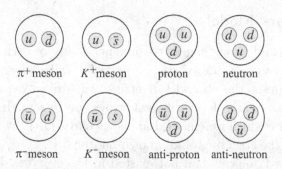

Fig. 8.2   Quark content of some mesons and baryons.

According to the quark model, all mesons are made of a quark and antiquark. For example, a positively charged pion denoted as π^+ comprises an up quark and a down antiquark. In other words, its *quark content* is $u\bar{d}$. A particle called *positive K meson*, or *positive kaon*,[3] is made of an up quark and a strange antiquark ($u\bar{s}$). It is denoted as K^+. All baryons consist of three quarks. The quark content of the proton is uud, while that of the neutron is udd. The quark content of antimesons and antibaryons is obtained by replacing each quark by the corresponding antiquark and vice versa. For example, the content of an antiproton is $\bar{u}\bar{u}\bar{d}$. The antiparticle of a positive pion is a negative pion ($d\bar{u}$), while that of a positive kaon is a negative kaon ($s\bar{u}$). The quark content of all these particles is presented in Fig. 8.2.

The quark model successfully explains the structure of all known hadrons using such simple combinations. Moreover, the model predicted new particles. Three distinct flavors of quarks can produce $3 \times 3 = 9$ different quark-antiquark combinations, and the particle corresponding to each of these combinations must exist. Similarly, there should exist $3 \times 3 \times 3 = 27$ different baryons made of three quarks.[4] When the quark model was formulated, many particles predicted by it were already known, but some of them had to be found.

---

[3]Here, the word 'positive' refers to the positive charge of kaon.

[4]In fact, the quark model predicts a larger number of hadrons. Additional particles are made of the same three quark flavors in various combinations but differ by some other intrinsic properties.

The discovery of each new particle evokes an unspeakable thunder of emotions, which can be compared to the depth of feelings experienced by a navigator when he discerns a tiny strip of unknown land on the horizon for the first time. For these rare moments of glory, physicists spend an infinite string of days and nights near their instruments and computers fixing faults or analyzing data. But after the advent of the quark model, the search for new particles began to resemble, more and more, a countryside walk by a signposted trail rather than the sailing of uncharted seas. Long before the actual discovery of each new particle, the corresponding combination of three quarks was already known, and the properties of the expected particle were already specified. Experimenters just needed to find the predicted object in the right place. Despite this, the enthusiasm of physicists did not waver. The discovery of each new meson or baryon solidified confidence in the quark model. The model has been confirmed over and over again and has finally been approved as the foundation for the description of hadron structure.

For all their complete happiness, scientists missed just one thing. Nobody could extract quarks from hadrons. In all other cases, the division of an object into constituent parts was always possible. In the atom, a nucleus can be easily separated from electrons. Similarly, individual protons and neutrons can be taken out of a nucleus. But releasing quarks from a hadron was never possible. There were attempts to produce free quarks in collisions of hadrons. However, these collisions created a lot of different mesons and baryons instead of quarks. Increasing the energy of collisions resulted in a growing number of particles produced, but quarks never appeared. Quarks were searched for at electron-positron colliders. The energy of collided electrons and positrons was more than sufficient for quark production. However, electron-positron annihilation always generated hadrons or leptons, but not quarks. These negative results led to the only possible conclusion: quarks cannot be separated from one another. For some reason, they are always confined inside mesons and baryons. After all failed attempts, this behavior of quarks was accepted as their inherent property, which remained to be understood.

Gradually, physicists started to treat quarks not only as the constituents of hadrons but as the main participants of the strong interaction. For example, the collision of two protons was considered to be the interaction of quarks contained in them. Similarly, the creation of hadrons in electron-positron annihilation was described as the production of a quark-antiquark pair, assuming that the pair was later converted into hadrons. Explaining such processes in terms of quarks turned out to be more convenient than doing so in terms of mesons and baryons. Besides, quarks revealed one important property, which became very helpful in building the theory of strong interaction. This property consists of the following.

It turns out that each quark exists in one of three possible *states*. This phrase may look enigmatic, but it is quite similar to saying that a particular person is present in one of the rooms of a house. In this case, the room where the person is currently situated can be called his or her state. To distinguish the states of quark, they are given the names of three colors: *red, green* and *blue*. For example, we can speak of a green up or a red strange quark. The states of antiquarks are called *cyan, magenta* and *yellow*, respectively. All these names don't imply that colors are somehow related to quarks. It is just a way to distinguish three states. In principle, instead of colors, we could speak of "an up quark in the state A", or "a strange quark in the state B", but using the common names seems more convenient. This is one more example where familiar words are employed to denote scientific terms. In addition to quarks, several other particles have similar states. Such objects, including quarks themselves, are called *color particles*. In the following discussion, the phrase "a particle has color" means that a particle has several possible *color states* and can be in one of them. Particles which don't have color states are called *white*.

It turns out that all mesons and baryons are white particles, although they are made of color quarks. The reason for this will become clear a little later. The requirement to be a white particle implies that out of nine possible combinations of colors of quark and antiquark, only three are allowed for a meson. These combinations are red and cyan, green and magenta, or blue and yellow. It is well known that the superposition of red and cyan beams of light gives a white spot on a screen. The same result is obtained by adding green and magenta or blue and yellow beams. A similar effect is produced in mesons. The colors of quark and antiquark cancel each other in the specified combinations, and the composite particle becomes white. Of course, this result is not evident, but it can be proven using mathematical formulae. It is also well-known that a blend of red, green and blue rays results in the color white. Similarly, each baryon has one red, one green and one blue quark and their combination makes the baryon white. Any other combination of three color quarks produces color objects. In its turn, the antibaryon is made of cyan, magenta and yellow antiquarks, and the mixture of these colors also creates a white particle. The presented rules of the addition of quark colors partially motivate the usage of the term "color" in the microworld of particles. In all other respects, the color of a beam of light and the color of a particle are entirely different phenomena. For example, the color of light can continuously vary from infra-red to ultra-violet. But the quarks can have just three predefined colors and absolutely no other shades.

Leptons, contrary to quarks, don't have color states. Using the previous analogy, we can say that a quark lives in a three-room house, while a lepton settles for a one-room studio. This difference between quarks and leptons may look like one more dull property of particles, but it, in fact, has far reaching consequences.

Quarks and leptons are very similar to each other. All of them are fermions. Hence, their behavior is described by the same mathematical equations. Essentially, there are just two major differences between quarks and leptons. **First**, quarks participate in the strong interaction while leptons do not. And **second**, quarks, contrary to leptons, have colors. From this observation, just a little step is needed to derive a fundamental conclusion. Relating these two differences, we can suppose that **quarks participate in the strong interaction because they are color particles**. Figuratively speaking, quarks interact strongly just because their houses have three color rooms. Leptons don't have color rooms and, therefore, are prevented from participating in the strong interaction. After all, such a differentiation is quite familiar to people. Balls, where a select few interact in a very special way, are always given in palaces and never in studios.

Thus, color may be that sought after property that allows particles to interact strongly. Conversely, the absence of color prevents particles from doing this. As it was stated before, the existence of such a property is essential for building a comprehensive theory of strong interaction.

The first idea on the role of quark color in strong interactions emerged at the end of the seventies of the last century. As the following developments demonstrated, this idea was absolutely right. Of course, initially it was just a bold hypothesis, and a lot had to be done to prove it. The specific relation between color and the strong interaction had to be established. Also, the fundamental boson of the strong interaction had to be identified. Still, realizing the importance of colors was a major break-through in understanding the strong interaction. One song, which was very popular at the time of creation of the quark theory, said: "I've found a driver and that's a start".[5] Similarly, with the discovery of color quarks, the main participants of the strong interaction have been found, and that has become an excellent start for the future development of the theory.

8.3   Gluon

The full-scale theory of strong interaction was formulated at the beginning of the eighties. Many theorists participated in its creation. In general, particle physics by that time became a collective science. Both theorists and experimenters worked in large collaborations. This tendency has been particularly typical for experiments, which initially involved dozens, then hundreds, and finally thousands of physicists. For this reason, each victory over the forces of nature has been rightfully regarded as a joint success of all those participants.

The theory of strong interaction is called *quantum chromodynamics*, or just *QCD*. In this name, the part "chromo" is derived from the Greek word meaning "color". The color indeed plays the central role in this theory. But once again, it should be stressed that color

---

[5]This is part of the song "Drive my car" by Beatles, which was released in 1965.

in particle physics is not related at all to the ordinary visual colors of our world.

Quarks are the main particles acting in the QCD. Like all other forces of nature, the strong interaction between quarks consists of exchanging the corresponding fundamental boson which is called the *gluon*. The word, "gluon", which is derived from "glue", reflects the role of this particle in binding together quarks inside hadrons. A gluon interacts only with color particles and does not notice white objects, such as leptons. Hence, the color can be considered as the charge of the strong interaction. Often, it is even called the *color charge*.

A quark can change its color after interacting with gluons. Using the previously mentioned analogy with a house, we can say that the gluon can move a quark from one color room to another. Otherwise, the strong interaction is quite similar to electromagnetic interaction. In particular, two gluons can create a pair of quark and antiquark. Inversely, a quark and antiquark can annihilate and produce gluons. Also, the strong interaction, like the electromagnetic one, can never change the flavor of a quark.

All hadrons are white particles. Nevertheless, they strongly interact because they are made of color quarks. To understand the mechanism of the strong interaction of hadrons, we can use the analogy with ordinary colors. Let us consider how color images are displayed on a computer screen. The screens of most monitors are composed of many pixels, and each pixel has red, green, and blue components called subpixels. The color of each subpixel is clearly visible with a close-up view. But at a typical distance between the human eye and the screen, the subpixels are not discerned, and their colors are blended. In particular, if the intensities of all three subpixels are equal, the visible color of the corresponding pixel is seen as white. A similar effect happens in the microworld of particles. When two hadrons are close enough, for example, when they collide or when they are both located inside the nucleus of an atom, the color quark in one hadron distinguishes the color quarks in another hadron and interacts with them by exchanging gluons. But when the distance between two hadrons increases, they gradually "perceive" each other as white particles. Hence, the exchange of gluons between them diminishes, the strong interaction decreases and, finally, completely vanishes. Therefore, a white hadron does not experience the strong force of other white hadrons if large distances separate them.

Using this observation, let us consider a nucleus once again. The distance between baryons in a nucleus is quite small. So, they strongly interact with each other. This interaction results in the mutual attraction of baryons. The attraction is so strong that most nuclei are very stable and are very difficult to break into pieces. On the contrary, the distance between the nuclei of neighbor atoms in any material is much larger. It is sufficient to recall the comparison of an atom with a football field. If the atom were of the size of a football field, the nucleus would have the size of a tennis ball in the center of the field. At such a huge distance, baryons of the neighbor nuclei perceive each other as white particles and do not exchange gluons. Consequently, nuclei do not attract each other and exist as separate objects.

So, the atomic structure of matter is fully explained by the sensitivity of gluons only to the color of particles and by the fact that all hadrons are white. Just this result alone can be considered to be a big success of QCD. But this is not the only achievement of this theory. It also helps to understand why all hadrons are white and all quarks are confined inside hadrons.[6] This happens because of one special feature of the strong interaction. It turns out that the strong force acting between two color objects does not decrease when they move apart. On the contrary, the force remains constant at large distances. Such a behavior is unique to the strong interaction. All other known forces of nature diminish with an increase in the distance between interacting objects. For example, the gravitational attraction of a star drops very quickly with an increase in a distance to it and becomes negligible when the distance is very large. But the strong force between two isolated quarks remains substantially the same even when the quarks are very far from one another.

Now, suppose that we try to extract a quark from a baryon. For example, we strike a baryon with an accelerated electron. What happens with the baryon after this collision is schematically

---

[6]In fact, the confinement of quarks is not yet analytically proven. However, the general expectation is that QCD causes this effect.

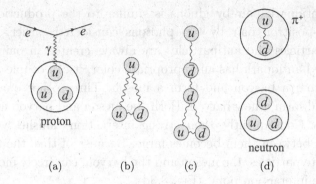

Fig. 8.3   Attempt of the quark to escape from a baryon.

shown in Fig. 8.3. The electron interacts electromagnetically with
an individual quark. The energy transferred from the electron to
the quark can be significant so that the quark indeed can start to
move away from two other quarks of the baryon (Fig. 8.3b). However,
the strong force continues to attract the quarks to each other. This
attraction remains the same even when the quarks move very far
apart. Moving against force always requires energy. When we pull
apart two objects connected by a spring, we spend some energy to
oppose the force exerted by the spring. Similarly, energy is needed to
separate the quarks attracted by the strong force. It turns out that
the required energy is enormous. For example, separating two quarks
by just one millimeter would require an amount of energy about five
million times greater than what can be achieved by accelerating particles in the most powerful existing collider. Thus, even though the
energy of collision can be very high, it can never be sufficient to pull
the quarks apart and set them free.

Nevertheless, the separation of quarks is impossible not just
because it would require enormous energy. In the case of the spring,
the energy applied to pull apart the objects transforms into the
energy of the spring. Similarly, the initial energy of the quark, which
it receives after the collision and which is used for quark separation,
converts into the energy of gluons acting between quarks. When the
energy accumulated by gluons becomes high enough, the gluons produce a pair of quark and antiquark (Fig. 8.3c). The creation of a

quark-antiquark pair by gluons is similar to the production of an electron-positron pair by two photons considered earlier. In both cases, particles and antiparticles are always created in pairs. If the generated antiquark has an appropriate color, it can couple with the quark from a baryon and form a meson. Three other quarks also unite and form a new baryon. Both the meson and baryon are white particles. Consequently, their strong interaction vanishes when the distance between them becomes large. Because of this, the two new composite particles, the meson and the baryon, can freely move away from the interaction point (Fig. 8.3d).

If the collision energy is high enough, gluons can produce several quark-antiquark pairs. But all these new quarks and antiquarks should also be bound together into mesons and baryons. That is why the rise in the collision energy increases the number of hadrons created. Some of these new particles contain quarks from the original baryon. However, the extracted quarks can never exist as isolated objects. The strong force never allows them to break away.

All other attempts to isolate quarks are also doomed to fail. For example, let us consider an imaginary case of a quark-antiquark pair produced near a black hole. The gravitational attraction of a black hole is so strong that even light emitted inside a certain boundary around the black hole cannot escape. Suppose that a quark and antiquark pair is produced at this boundary. One of the quarks falls into the black hole, while the other can get away.[7] However, even in this case, the free quarks are not produced. When the distance between the quark and antiquark increases, the energy of the strong force between them becomes sufficient to create multiple quark-antiquark pairs at the boundary. When the colors of some of the produced quarks and antiquarks match, they will form white particles, which will be separated from other objects without much effort. Other quarks and antiquarks will continue to interact strongly and produce quark-antiquark pairs. The production of such pairs will continue until all quarks and antiquarks unite into white particles. Some of

---

[7]The emission of particles by black holes was first studied by the English theorist Stephen Hawking.

the white particles created will fall into the black hole, others will escape, but no color object will be emitted.

In all other cases, the energy applied to separate quarks also causes the creation of quark-antiquark pairs. The quarks and antiquarks produced always group and form mesons and baryons. All particles composed of quarks must be white because only in this case, they can be isolated from one another. The color objects made of quarks would continue to interact with the same force even if they were in different galaxies. Therefore, only white hadrons can exist as isolated particles, while quarks are forever confined inside them.

Let us draw some conclusions. In the previous chapter, we defined several stages of baryogenesis, which need to be elucidated. The development of the QCD helps to clarify some of them. This theory successfully explains why primordial quarks, which were created soon after the Big Bang, very quickly combined into mesons and · baryons. The QCD also makes clear why baryons were later united into nuclei. The additional input from the electromagnetic theory helps to understand how nuclei and electrons formed atoms. Hence, this part of baryogenesis becomes substantially clear.

Still, the QCD does not give a clue to the reason for antimatter disappearance. It's because the strong interaction is somewhat analogous to the electromagnetic interaction. In particular, both follow the same rule of pairs. Gluons, like photons, always create quarks and antiquarks in pairs. Moreover, the interaction of the gluon with quarks and antiquarks is essentially the same. Of course, the colors of a quark and antiquark are different, but this difference is like the opposite charges of the electron and positron. It does not make the behavior of quarks and antiquarks dissimilar in the strong interaction. Thus, according to QCD, the number of baryons formed by quarks should be equal to the number of antibaryons containing antiquarks, provided the initial number of quarks and antiquarks was the same. This conclusion means that the search to explain the mystery of missing antimatter must continue by studying other types of interaction.

Chapter 9

Decays of Quarks and Leptons

9.1 Weak Interaction

Primordial matter of the newborn universe was created in the form of fundamental fermions, that is, leptons and quarks. All twelve types of these particles were produced equally well. In this respect, nature was quite liberal and did not grant any preference to a particular flavor. After that, the process of baryogenesis began. Quarks were combined into mesons and baryons under the influence of the strong interaction. However, most quarks did not survive even within hadrons. All mesons quickly disintegrated. Their disappearance was caused by the decays of the corresponding quarks. Almost all baryons did not exist for a long time for the same reason. Only protons and neutrons bound inside nuclei could remain unchanged from the beginning of time to our days.

Protons and neutrons contain only up and down quarks. So, just these two quark flavors are present in all known materials. Four other types (strange, charm, beauty and top) have completely disappeared. Physicists know how to create them artificially in collider experiments. Nonetheless, today, like in the early universe, particles of these unstable flavors exist only for a very short time. After that, they *change flavor* and turn into up or down quarks. Similarly, two lepton flavors, muon and tauon, convert into corresponding neutrinos.

The change of particle flavor is an extraordinary effect, which is hard to compare to any other phenomena. The transformation of one substance into another is well-known. For example, liquid water can be obtained from two gases, hydrogen and oxygen. But this conversion just means that a molecule of water contains atoms of these two chemical elements. A table can also be assembled from several pieces of wood or other material, and this does not seem unusual at all. The transformation of particles is something completely different. It is more like a miracle from fairy tales, where a pumpkin suddenly turns into a golden carriage and mice become horses. All of us have grown up on fairy tales, and in childhood, we were ready to believe that a nutcracker can be transformed into a handsome prince. After becoming adults, we convinced ourselves that such miracles cannot happen. But in the microworld of particles, everything is possible. One particle can turn into another. Moreover, such transformations are standard and frequent there.

A top quark is subject to the longest series of changes. This heaviest known particle transforms into a bottom quark, whose mass is almost forty times smaller. If an enchanter decided to do a similar trick with animals, he would need to turn a horse into a poodle. The bottom quark becomes a charm quark, which, after some time, transmutes into a strange quark. At last, the strange quark turns into an up quark, and the cascade of wonderful conversions finally ends.

Most of the fundamental fermions, which were created in the first few instants after the universe's birth, changed their flavor and were finally transformed into up or down quarks, electrons or the three types of neutrinos. Hence, the processes of flavor change constitute an important part of baryogenesis. They can also be essential for understanding the absence of antimatter in the contemporary universe, just because they cause the disappearance of certain particles and creation of other particles instead. That is why the rules determining the flavor change of fundamental fermions are worth considering in more detail.

For a long time, scientists did not suspect the existence of flavors and the ability of particles to transform from one type to another. In fact, in 1932, physicists enjoyed a very brief period of calm and contentment. At that time, they thought that all particles existing in nature had been found and that the structure of matter had been completely understood. That year, the English scientist James Chadwick discovered a neutron. With this discovery, all particles contained in atoms became known. A proton, neutron, and electron were thought to be entirely sufficient to explain the structure of the atom, and no more additional particles, except probably a mysterious neutrino, were required.

But the happiness of scientists was fleeting. In the same year of 1932, Carl Anderson discovered a positron (see Section 3.4). Four years later, he and his student Seth Neddermeyer found one more particle named the *muon*.[1] The muon did not fit at all into the neat and clear description of nature, which had been established by that time. It did not have anything to do with atoms. That is why it was initially thought of as an "unnecessary" particle. The perplexed feelings of scientists were nicely reflected in the famous phrase about a muon said by the Nobel prize winner Isidor Rabi: "Who ordered this?".[2] Understanding of the importance of the muon came much later when this particle took its rightful place among the twelve fundamental fermions representing the foundation of matter.

One of the most unusual properties of the muon consists of its brief lifetime. Soon after its creation, the muon ceases to exist. Instead, three new particles appear: an electron and a neutrino-antineutrino pair. The disappearance of the muon is called *decay*, but this term does not mean that a muon is made of an electron, neutrino and antineutrino and that its constituent particles are released when a muon decays. It is the transformation of one object into three other. Such changes exist only in the microworld of particles. Elsewhere, they are called miracles.

Like in β decay considered earlier, the only detectable particle in muon decay is the electron. The production of neutrinos is identified by the difference in energy of the initial muon and the final electron. But in contrast to β decay, the remaining energy in muon decay is taken away by not just one, but two invisible particles.

---

[1] Initially this particle was called μ meson, but it soon became evident that the muon was not related to mesons. Hence, its name was modified.

[2] The story goes that he said this phrase at a dinner with friends at a Chinese restaurant when they were served a dish which nobody wanted.

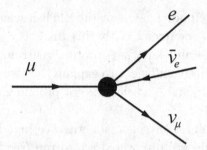

Fig. 9.1   Diagram of muon decay within the theory by Fermi.

β and muon decays have a lot in common because both of them are governed by *weak interaction*. The first theoretical description of weak interaction was proposed by the Italian scientist Enrico Fermi in 1933. Initially, Fermi developed his theory to describe only β decay, since at that time the muon had not yet been discovered. But it turned out that this theory of Fermi's worked equally well in the case of a muon. In this theory, muon decay was presented by the diagram shown in Fig. 9.1. At that time, essentially nothing was known about the weak interaction. That is why the details of the physical process, which happened at the point where four fermion legs crossed, were hidden by a large black blob. Notwithstanding this limitation, the theory worked pretty well in the case of muon decay and many other similar processes.

Except for the black blob, the diagram in Fig. 9.1 follows the same general rules of all Feynman diagrams. Since it describes the process of decay, the initial state contains just one particle, a muon, which is denoted by the Greek letter μ. The final state includes an electron, a muon neutrino (denoted as ν_μ) and an electron antineutrino ($\bar{\nu}_e$). Like in all other diagrams with antiparticles, the arrow on the $\bar{\nu}_e$ leg is opposite to the direction of time flow.

The Fermi theory was used in particle physics for more than 30 years. But in 1968, a new theory of weak interaction emerged. It had been developed by Sheldon Glashow, Abdus Salam, and Steven Weinberg. Soon after its invention, the theory was called the *Standard Model* of particle physics. This title implies that the proposed theory has become the baseline model describing both weak and electromagnetic interactions. Indeed, the Standard Model has explained many known phenomena in particle physics and has predicted many new effects. Its predictions are continuously tested experimentally, and in all cases, the theoretical expectations agree very well with experimental results. The Standard Model not only elucidates the properties of weak interaction, but it also unifies the weak and electromagnetic forces and presents them as two related parts of the *electroweak interaction*. According to this theory, both forces of nature have a common origin and are therefore described by similar mathematical formulae.

The success of the Standard Model is tremendous, and the full list of its achievements is very long. Thus, it is not surprising that Glashow, Salam, and Weinberg have been awarded the Nobel Prize. One of the most significant results of this theory is the explication of processes that involve the change of flavor. The Fermi theory treated the transformations of quarks and leptons as an empirical fact, without properly understanding why it happens. The mechanism of this fascinating phenomenon became clear only after the development of the Standard Model. Some of the main features of this mechanism, as revealed by the Standard Model, are discussed in the next section.

9.2   Flavor Transformations

The Standard Model includes all essential parts required for a comprehensive description of weak interaction. In particular, it introduces two new fundamental bosons responsible for the weak force. They are named the W and Z *bosons*. Weak interaction represents the exchange of these bosons between fermions, so its general structure is similar to that of the electromagnetic and strong interactions.

The fundamental bosons of all interactions are unique in one way or another. The special feature of the W and Z bosons is that both of them have mass. Furthermore, these bosons are much heavier than most fermions. Because of the large mass of these bosons, the weak interaction is indeed weak, fully justifying its name. This weakness is quite easy to understand by recalling the football analogy of boson exchange mentioned earlier. Indeed, a football game would not be too dynamic if its participants were requested to pass each other an iron ball. Something similar happens in the weak interaction. Particles exchange very massive bosons, and the high mass of bosons weakens their interaction.

In addition to the mass, the W boson has a charge which is equal in magnitude to that of an electron. In fact, there are two particles, W^+ and W^-, which are charged positively and negatively, respectively. All charged particles interact electromagnetically, and so does the W boson. In contrast, the Z boson is a neutral particle, and it does not couple with a photon. Going forward, these properties of the W and Z bosons fully determine the rules of flavor change of all fundamental fermions.

The elementary vertices of the weak interaction of fermions with the W boson are very similar to those of electromagnetic interaction. For example, the vertex transforming a muon into a muon neutrino, which is shown in Fig. 9.2 (left), reminds us of the vertex with a photon presented in Fig. 4.2. This similarity can be expected since both interactions have a common origin and are described within the same theoretical framework. But there is also a significant difference between the two interactions. The W boson changes the flavor of fermions, while the photon never does this. Since the W boson is a charged particle, and the total charge should always be conserved, the interaction with the W boson changes the charge of the participating fermion by one unit.[3] The modification of fermion charge is equivalent to the change

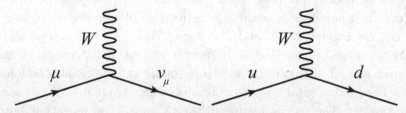

Fig. 9.2 Vertex of the weak interaction transforming a muon into a muon neutrino (left) and an up quark into a down quark (right).

---

[3]The charge of all particles is always determined relative to the magnitude of the electron charge. In these units, the charge of W boson can be either $+1$ or -1.

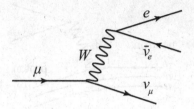

Fig. 9.3 Diagram of muon decay within the Standard Model.

of its flavor. That is why the interaction with a W boson always results in the conversion of one fermion into another. In the example of Fig. 9.2 (left), the muon with charge -1 becomes a muon neutrino with charge zero. In the same way, an up quark (the charge $+\frac{2}{3}$) after interaction with a W boson is transformed into a down quark (the charge $-\frac{1}{3}$) as shown in Fig. 9.2 (right).

The diagram of any process involving a W boson can be obtained by combining several W vertices. For example, muon decay is presented by the diagram shown in Fig. 9.3. This process includes two independent transformations. In one vertex, the muon turns into a muon neutrino. This *transition* can be denoted as $\mu \rightarrow \nu_\mu$. In the second vertex, the electron neutrino becomes an electron ($\nu_e \rightarrow e$). Of course, it can be argued that the actual process happening in this vertex consists of the simultaneous creation of an electron and electron antineutrino. However, from the theoretical point of view, **the type of flavor transformation is defined by the direction of the arrows on the fermion lines.** So, in the case of Fig. 9.3, it is described as the transition of a neutrino to an electron. Since the arrow on the electron neutrino leg is opposite to the flow of time, this leg corresponds to an antiparticle. Nevertheless, the underlying transition is $\nu_e \rightarrow e$. Similarly, in all other cases, the type of transition is unambiguously determined by the direction of arrows on the fermion lines.

The weak interaction also effects β decay. In this process, a neutron turns into a proton with the additional production of an electron and an electron antineutrino. At the quark level, the conversion of a neutron to a proton corresponds to the transformation of one of the down quarks, which is contained in the neutron, into an up quark,

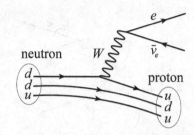

Fig. 9.4   Diagram of β decay within the Standard Model.

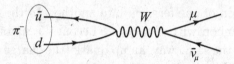

Fig. 9.5   Diagram of π^- decay.

which is included in the proton. Two other quarks of the neutron do not participate in the interaction and are just passively inherited by the proton. That is why they are called *spectator quarks*. The diagram corresponding to β decay is presented in Fig. 9.4. Like in muon decay, β decay consists of two transitions: $d \to u$ and $\nu_e \to e$. In the first transition, the charge of the fermion increases by one, and in the second it decreases by the same amount. Hence, the total charge is conserved.

Flavor change is responsible for the decays of many hadrons. As a typical example of such a process, let us consider the decay of a negatively charged pion to a muon and muon antineutrino. The corresponding diagram is shown in Fig. 9.5. The π^- meson is made of a down quark and an up antiquark. The quark and antiquark of the π^- meson annihilate and produce a pair comprising of a lepton and an antilepton. The presented diagram is quite different from that in Fig. 9.4, but the underlying weak interaction is essentially the same, except that the leg corresponding to the u quark is directed opposite to the time flow, and the transition $\nu_e \to e$ in β decay is replaced by the transition $\nu_\mu \to \mu$.

Instead of a lepton and neutrino, an additional pair of quarks can participate in the weak interaction. For example, the diagram

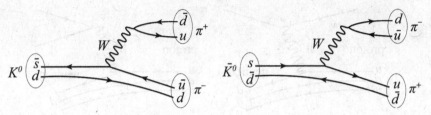

Fig. 9.6 Diagrams of the K^0 (left) and $\bar{K}^0$ (right) decay to π^+ and π^- mesons.

of the decay of a *neutral kaon*, which is denoted as K^0, is shown in Fig. 9.6 (left). The K^0 meson contains a down quark and a strange antiquark,[4] that is, its quark content is $d\bar{s}$. The diagram shows that in the K^0 decay, the strange antiquark becomes the up antiquark. However, we should always follow the arrows on the fermion lines to determine the type of flavor change. Thus, in this decay, the actual transition is $u \to s$. Similarly, another flavor change in this decay is $d \to u$. The quarks and antiquarks produced in the K^0 decay cannot exist as isolated particles. Consequently, they are united by the strong interaction into π^+ and π^- mesons. Hence, these two particles are observed in the final state. It is important to note that the $\bar{K}^0$ meson, which is an antiparticle of K^0 meson with quark content $s\bar{d}$, also decays into a pair of π^+ and π^- mesons. The corresponding diagram is shown in Fig. 9.6 (right).

All examples mentioned illustrate the main rules of the flavor change in weak interactions. They are as follows. **First**, each process of flavor change involves four fundamental fermions. They can be four leptons (Fig. 9.3), two quarks and two leptons (Figs. 9.4 and 9.5), or four quarks (Fig. 9.6). **Second**, four participants are divided into two pairs. Each pair independently interacts with the W boson. The pair contains either two quarks or two leptons but never a quark and a lepton. **Third**, in each pair, the transformation of one fermion into another causes the change of charge by one unit. This rule explains, in

---

[4]By convention, a meson containing s quark is called an antiparticle, and a meson with strange antiquark is called a particle. But this convention is entirely arbitrary. Each meson contains a quark and antiquark, and we cannot specify which meson is a particle and which is antiparticle.

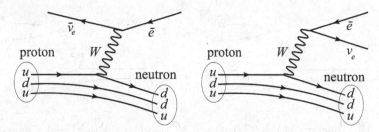

Fig. 9.7   Diagram of the process used to discover antineutrino (left). Diagram of positron emission (right).

particular, why the conversion of a quark into a lepton is forbidden.[5] In one pair, the charge always increases by one unit, and in another, it decreases by one unit, so that the total charge remains the same.

We also need to mention the **fourth rule**, which will be considered in more detail later in Section 15.2. The so-called *horizontal transitions* between the fermions having the same charge, such as $d \to s$ or $\mu \to e$ are forbidden. These transitions could occur after interaction with either a photon or a Z boson.[6] However, the rules of the Standard Model explicitly prohibit the change of flavor without a change in the fermion charge.

All effects of weak interaction can be explained by these rules. For example, the process used to discover the antineutrino in the experiment of Cowan and Reines (see Section 6.1) is shown in Fig. 9.7 (left). This process consists of two independent flavor changes: $u \to d$ and $e \to \nu_e$. The arrows on the legs corresponding to the election and neutrino are directed opposite to the time flow. Therefore, both these legs correspond to antiparticles.

The rules of weak interaction remain the same for the processes involving other flavors of quarks and leptons. Quarks are color particles and are always confined inside hadrons. Consequently, their transformation to other quarks leads to the decay of the corresponding hadron. For example, a particle called D^+ meson contains a

---

[5]In the hypothetical transition from quark to lepton, the change of charge can never be equal to one unit.

[6]Both these bosons are neutral particles and, consequently, cannot change the charge of the interacting fermion.

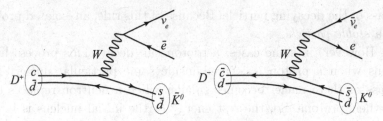

Fig. 9.8   Left: Diagram of decay of D^+ meson into an antikaon, a positron and electron neutrino. Right: Similar decay of D^- meson.

charm quark and a down antiquark. The charm quark can transform into a strange quark and, also, produce a positron and neutrino. The corresponding diagram is shown in Fig. 9.8 (left). The created strange quark (s) unites with the down antiquark $(\bar{d})$, which is left after D^+ disintegration. These two objects form a $\bar{K}^0$ meson. In the next step, the $\bar{K}^0$ meson decays as per the diagram in Fig. 9.6 and creates a pair of positive and negative pions. Both these pions disintegrate as shown in the diagram of Fig. 9.5. The muons, which appear after pion disintegrations, also decay, as can be seen in the diagram of Fig. 9.3. Thus, the decay of a D^+ meson finally leaves nothing but electrons, positrons, neutrinos, and antineutrinos. Most other hadrons undergo similar processes and also disappear. In fact, only a proton can exist for an infinitely long period of time.

The decay of a proton would be possible if one of its u quarks were transformed into a d quark, as in the diagram in Fig. 9.7 (right). The created d quark would combine with the remaining u and d quarks and would form a neutron. The diagram in Fig. 9.7 (right) can be obtained from the diagram in Fig. 9.7 (left) by rotating the neutrino line. Therefore, in principle, the corresponding process could exist. However, the mass of a neutron is greater than the mass of a proton. Consequently, the energy of a neutron, which always includes its rest energy, exceeds the rest energy of a proton. Thus, the decay of a proton would correspond to the transition from a state with lower energy to a state with higher energy. Such a transition is forbidden by the law of energy conservation, which can never be violated. The example considered is a particular case of a general rule, which states that the sum of masses of the decay products can never exceed the mass of the decaying particle. Because of this rule, an isolated proton is a *stable particle*.

However, in some cases, a proton can decay. This process happens when a proton is located inside some particular nuclei. The mass of these nuclei becomes smaller when a neutron replaces one of their protons. So, the rest energy of the initial nucleus is larger than the rest energy of the new nucleus. Consequently, the transition corresponding to the diagram in Fig. 9.7 (right) becomes possible. The excess energy is transferred to the positron and neutrino produced, so that total energy is conserved, as always. Such a process is called *positron emission*. Nuclei capable of positron emission are used in the method of positron-emission tomography discussed earlier in Section 5.2.

Contrary to a proton, an isolated neutron is an *unstable* particle. It decays within a few minutes after its production, as shown in the diagram of Fig. 9.4. But the decay of the neutron contained inside a nucleus is prevented by the same law of energy conservation. In most cases, a nucleus with a neutron has a smaller mass than a modified nucleus with the neutron replaced by a proton. The corresponding deficit of energy does not allow the neutron to decay. That is why most nuclei can exist for a very long time, and this is exactly what they do for many billions of years. Only in rare cases, the replacement of a neutron by a proton decreases the mass of a nucleus. Consequently, such nuclei are unstable and undergo β decay. The corresponding materials are used as fuel in nuclear power stations.

Let us summarize the results of this section. Weak interaction is the only force of nature that is responsible for the change of flavor of all quarks and leptons. The rules of flavor transformations are clearly defined by the Standard Model. Because of these rules, most hadrons are unstable particles. Hence, the theory of weak interactions fully explains what happens with primary quarks and leptons created soon after the Big Bang. Moreover, it precisely describes each step of their transformation. As a result, the knowledge gained of electromagnetic, strong and weak forces completely clarifies a considerable part of baryogenesis. In a few words, it can be depicted as follows.

The primordial matter of the universe was a mixture of quarks and leptons. When the universe cooled enough and primary radiation could not produce new fermions anymore, all quarks were combined into hadrons by the action of the strong force. Later, most hadrons decayed because the weak interaction caused the flavor change of almost all quarks. The only surviving hadrons were protons and neutrons. Protons had no opportunity to decay, while neutrons survived because they had some time to form nuclei together with protons. The weak interaction was also responsible for the decays of all muons and tauons produced at the initial stages of matter formation. After the decays of these leptons, only electrons and three flavors of neutrino remained.

While the picture presented is clear enough, it is not complete, because it does not explain what has happened with antiquarks. At first glance, they should be able to follow the same chains of interactions and form antinuclei, which can also exist forever. Indeed, the strong interaction of antiquarks is the same, and the change of antiquark flavors seems identical to that of quarks. For example, the diagram of the transformation of a charm antiquark shown in Fig. 9.8 (right) looks very similar to the diagram for a charm quark in Fig. 9.8 (left). The same conclusion can be made on the transformations of all other particles and the corresponding antiparticles. However, there is a very subtle but significant difference between the processes involving particles and antiparticles. As is well known, the devil is in the details. In this case, the details decide the fate of antiparticles, in particular and that of the universe, in general. Indeed, a more thorough study leads to a surprising result: the symmetry between particles and antiparticles is really broken in weak interactions. In its turn, the broken symmetry has caused complete disappearance of antimatter. But before explaining the tiny difference between particles and antiparticles, which is crucial for the whole universe, we need to analyze more thoroughly what the term 'symmetry' actually means in physics. This attention to detail will be richly rewarded. It will lead to the discovery of one of the most amazing properties of nature and will become the first important success in the campaign to crack the mystery of the vanished antimatter.

Chapter 10

Symmetry and Particles

10.1 Mirror Reflection

The most straightforward way to investigate differences between particles and antiparticles is to consider a *symmetry* between them. Usually, the symmetry is defined as a similarity or exact correspondence of various objects or natural phenomena. In its conventional meaning, the word "symmetry" reflects a harmony of the world, an external or internal beauty of nature, people or their creations. The perfection of many buildings, such as churches and palaces, is achieved by an explicit or hidden symmetry of their shape, and the beautiful and attractive appearance of humans is explained to a large extent by the symmetry of our body. The symmetry of many constructions gives them the necessary strength for functionality while simultaneously converting them from purely utilitarian structures into a source of admiration. For example, a particular symmetry of the Eiffel Tower, which ensures an equal distribution of the weight among its four legs, makes the tower very stable and transforms it into one of the most admired landmarks around the world.

In physics, symmetry is related to the properties of a *physical system*. A physical system is just an object or a group of objects which are chosen for analysis. The symmetry of a physical system means that the behavior of the system remains the same even when some conditions are changed. For example, the spherical shape of a football guarantees that the rebound of the ball remains the same regardless of the side with which it hits the playground or the head of a player. FIFA imposes stringent requirements on the roundness of the ball. The variation of the diameter of the balls used in the World Cups must not exceed 1.5%. Thus, to a good approximation, the ball has *rotational symmetry*. This term means that the properties of the ball, and in particular its rebound, do not change under any rotation. The football pitch must also satisfy some requirements of symmetry. It should be reasonably flat, without bumps and hollows. In this case, the pitch can be treated as having *translational symmetry*. This type of symmetry assures the same rebound of the football in all parts of the pitch.

The concept of symmetry helps in solving many practical tasks. For example, the special symmetry of an airplane considerably simplifies the control of its flight. If we ignore the differences in the internal structure and the inscriptions on the surface of the plane, its left side looks very similar to its right side. This similarity is so strong that we would probably not be able to distinguish between the airplane and its reflection in a mirror. In other words, the airplane has *reflectional symmetry*. Because of this symmetry, the engines on the wings of the aircraft push it with the same force, and the flow of air acts in the same way on its left and right sides. This symmetry results in the flight of an airplane in a straight line without any additional attention from the pilots.

In particle physics, symmetry plays a central role. It is an essential part of the theoretical descriptions of all interactions. It allows scientists to understand various properties of known particles and predict the existence of new objects. For example, the quark model, which successfully describes the structure of hadrons, was formulated after observing and studying the symmetry in the behavior of hadrons. The Standard Model is also based on the laws of symmetry. Understanding these laws has led to the prediction of the existence of W and Z bosons. Symmetry is also important for explaining the mystery of the disappearance of antimatter.

The symmetry of an interaction means that the properties of the interaction do not change when there are variations of certain conditions. Some types of symmetry are relevant to all interactions.

For example, we can expect that particles interact in the same way in all points of space. Consequently, all interactions have translational symmetry. Also, all interactions are rotationally symmetric, since the rotation of all particles by the same angle should not influence the outcome of their interaction. However, some other types of symmetry are not universal. While being valid in some interactions, they are violated in others.

The *violation of a symmetry* means that the properties of a physical system change after variations of the corresponding conditions. The violation of symmetry significantly changes the behavior of a physical system and gives it entirely new properties. This effect can be understood by again using the example of an airplane. The structure of its wings includes special hinged surfaces called ailerons. The ailerons are located on the edges of the wings and are used for changing the direction of the airplane's flight. When the aileron on one wing moves upward, the similar aileron on the opposite wing moves downward. This change of the wing's geometry results in the violation of the reflectional symmetry of the airplane. The aircraft and its reflection do not look the same anymore. Because of this violation, aerodynamic force acts differently on the wings and turns the airplane about its longitudinal axis. As a result, the direction of the flight is changed. Thus, the aircraft with broken reflectional symmetry can perform new maneuvers. Similarly, an interaction that violates some symmetry acquires new qualities. This effect will be discussed later in this chapter, but before doing this, we need to consider one more important property of particles, which is called *chirality*.

In our everyday world, an object is called *chiral* if it is distinguishable from its mirror image. Screws, which are widely used in construction and technology, are a well-known example of a chiral object. The screw and its mirror image seem quite similar. Still, there is one important difference between them. They have opposite thread directions. The thread direction is one of the parameters of the screw which is as important as the screw diameter or length. But contrary to all other parameters, which can take many different values, the thread direction can be of just two types. It can be either right-handed or left-handed. The screw with the right-handed thread is tightened when it is turned clockwise, while the screw with the left-handed thread does the opposite. Apparently, there is no advantage between either thread direction. Nonetheless, almost all screws are right-handed, while the applications of the left-hand screws are limited. The thread direction of a screw can be called its chirality.

In the microworld, fermions also have the property of chirality. More precisely, each fermion can be in one of two *chiral states*. The meaning of the term "chiral state of a fermion" can be compared to the "color state of a quark". For both these terms, the analogy with different rooms of a house works well. The color of a quark is relevant to strong interaction. In turn, the chirality of a fermion is important for weak interaction.

The two chirality states of particles are named similarly to that of screws. The particles can be either *right-handed* or *left-handed* (Fig. 10.1). As in all other cases, when common words, such as "color" or "flavor", are used in the microworld, the states of particle chirality have a superficial association with thread directions. Let us try to imagine a spinning particle such that its rotation axis aligns with its direction of motion. In this case it is assumed that the left-handed particle spins counterclockwise and the right-handed particle spins clockwise. Strictly speaking, this comparison is not correct, because the classical concept of rotation is not applicable to particles.[1]

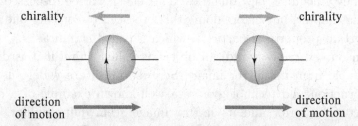

Fig. 10.1   Definition of the classical left-handed (left) and right-handed (right) particle.

---

[1] All fundamental fermions are treated as point-like objects, and they, therefore, cannot spin in the usual meaning of this term.

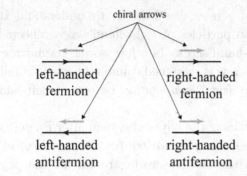

Fig. 10.2   Definition of chirality arrows in Feynman diagrams.

However, in many cases, the analogy between thread direction and the chirality of particles is acceptable and even quite convenient for understanding the effects related to chirality. Having said this, we also need to mention one important difference between thread direction and the chirality of particles. The thread direction of every screw is fixed. In contrast, particles can change their chirality. For example, a left-handed electron can become right-handed and vice versa.

The type of particles' chirality influences their interactions. More precisely, interaction of right-handed and left-handed particles can produce different results. To reflect this relation between chirality and interaction, chirality is sometimes indicated in Feynman diagrams. It is denoted as an additional arrow near the fermion line as shown in Fig. 10.2. The meaning of chiral arrows is simple to remember. Whenever this arrow is directed to the left, the corresponding particle or antiparticle is left-handed and vice versa. Note that the direction of the chiral arrows for particle and antiparticle are the same, while the directions of arrows on the fermion lines are opposite. We also want to remind the reader that the direction of the fermion arrows is not related to their direction of motion. It is a way to distinguish a particle from an antiparticle.

The chirality of particles is related to a special transformation of the physical system called the *operation of mirror reflection* or *parity conjugation*. Later, the term "operation" will often be used and will always mean some transformation of the physical system. The operation of parity conjugation is also denoted as *P-conjugation*.

The example of screws can help us to understand the action of P-conjugation on particles. A right-hand screw reflected in a mirror looks like a left-hand screw. In other words, the mirror reflection of the right-hand screw is tightened counterclockwise. Similarly, a right-handed particle after P-conjugation becomes a left-handed particle and vice versa.

If the properties of the physical system after P-conjugation do not change, then the *parity symmetry* (or *P-symmetry*) of this system is *conserved*. In the familiar world around us, the physical system and its mirror reflection always have the same properties. Simply speaking, a left-hand screw should be as strong as a similar right-hand screw, and, in general, traffic in the left-hand lane is as safe as in the right-hand one. But still, *violation of P-symmetry* often takes place in our life, especially in the case of traffic, see Fig. 10.3. In the microworld, *P violation* can also occur. This happens whenever the interactions of right-handed and left-handed fermions are different.

For a long time, scientists thought that the violation of P-symmetry was impossible. Indeed, numerous experiments with nuclei and atoms demonstrated that the electromagnetic and strong interactions of the left-handed and right-handed fermions are always the same. By inertia, physicists supposed that weak interaction also respects P-symmetry. However, this was incorrect. Weak interaction has many unique properties, making it a special force of nature. The behavior of weak interaction under parity conjugation demonstrates this distinction to its full extent.

One of the most unusual and not well-understood properties of weak interaction is the experimentally confirmed fact that only left-handed fermions and right-handed antifermions can participate in the W-boson exchange. For example, a decaying muon is always left-handed, and it always transforms into a left-handed muon neutrino. Also, a left-handed electron and a right-handed electron antineutrino are produced in this decay. The corresponding diagram, which takes into account chirality states, is shown in Fig. 10.4 (left). In the decay of an antimuon, fermions are also left-handed, and antifermions are right-handed as shown in Fig. 10.4 (right). Similar behavior is observed in all interactions with the W boson. The chirality

Fig. 10.3   Example of P-violation in the normal world.

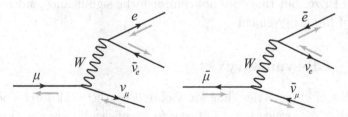

Fig. 10.4   Decay of muon (left) and antimuon (right) with chirality states of particles taken into account. Grey arrows near the fermion lines show the chirality of particles.

restriction does not mean that right-handed fermions exist for-ever. A right-handed muon can change its chirality and become left-handed. After that, nothing can prevent the muon from decaying.

Thus, weak interaction of left-handed and right-handed particles is different. To be more precise, right-handed particles do not interact with W bosons at all. This difference means that P-symmetry is violated in weak interactions. In other words, the mirror reflection of the microworld has completely different properties. Indeed, a left-handed muon can decay, while its mirror reflection, a right-handed muon can never do this. This behavior in the world we are familiar with would mean that only right-hand screws can be used to fasten components together, while the utilization of left-hand screws would be strictly forbidden. Of course, there is no such restriction for screws. A more frequent employment of right-hand screws is just the result of a multi-millennia tradition. But it turns out that weak interaction firmly obeys this law.

The violation of P-symmetry in weak interaction was discovered in 1957 by a group of American scientists directed by Chien-Shiung Wu, who was respectfully called "The First Lady of Physics." The result obtained was unexpected at that time and caused quite a stir. The conviction that P-symmetry could never be violated was firmly held, as is nicely demonstrated by the famous phrase by Pauli: "I refuse to believe that God is a weak left-hander".[2] He stated this just a few days before the announcement of the discovery of Wu. But after the results of Wu's experiment were published, the left-handed nature of the universe had to be accepted by everybody, including Pauli. Surprisingly, the discovery of Wu was not rewarded with the Nobel Prize, but this does not diminish the significance and recogni-tion of her achievement.

10.2   CP-symmetry

The list of symmetries that are violated by the weak force does not end at mirror reflection. Contrary to its misleading name, the weak

---

[2]He wrote this sentence in a letter to Victor Weisskopf.

C-conjugation

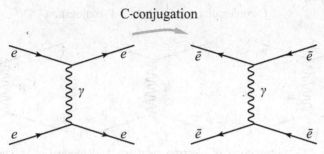

Fig. 10.5 The electromagnetic scattering of two electrons (left). Application of C-conjugation transforms this process into the electromagnetic scattering of two positrons (right).

interaction breaks many rules. One more type of symmetry, which is conserved in all interactions except the weak one, is called *charge symmetry*, or just *C-symmetry*. It corresponds to the operation of *charge (C) conjugation*. This operation replaces all particles in a particular process by corresponding antiparticles. If the behavior of particles in the process differs from the behavior of antiparticles in the C-conjugated process, the C-symmetry is violated. Otherwise, it is conserved.

After applying the operation of C-conjugation, the electromagnetic interaction of two electrons shown in Fig. 10.5 (left) becomes the interaction of two positrons (Fig. 10.5 (right)). It can be noted that the two diagrams can be obtained from one another by rotation by 180°. This observation helps to establish the conservation of C-symmetry in the displayed processes.

It turns out that the symmetry related to a given operation is always conserved whenever the conjugated Feynman diagrams can be obtained from each other by some rotation of the fermion lines. This statement is valid for all interactions and all types of symmetry. This rule, where applied to the process in Fig. 10.5, implies that charge symmetry is conserved in the electromagnetic interaction of electrons. Using a similar reasoning, we can conclude that charge symmetry must be conserved in the electromagnetic interactions of all charged particles.

In contrast to this, the weak interaction violates charge symmetry. This conclusion can also be derived using Feynman diagrams.

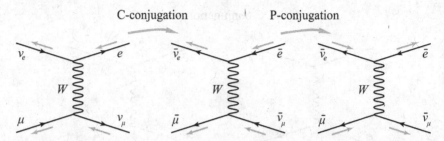

Fig. 10.6   The interaction of electron neutrino and muon (left). The process obtained after applying C-conjugation (center). The process obtained after applying C- and P-conjugations (right).

For example, let us apply the operation of charge conjugation to the interaction of an electron neutrino and a muon shown in Fig. 10.6 (left). After exchanging the W boson, these two particles are transformed into an electron and a muon neutrino. All particles participating in this interaction are left-handed. The diagram obtained after C-conjugation is shown in Fig. 10.6 (center). In this diagram, all left-handed particles are replaced by left-handed antiparticles. But left-handed antiparticles cannot interact with the W boson. Only right-handed antiparticles and left-handed particles can do this. Thus, C-conjugation results in a process that does not exist. This means that the charge symmetry is violated in weak interactions.

The violation of C-symmetry is a direct consequence of the discovery of Wu that "God is a weak left-hander", that is, that only left-handed particles and right-handed antiparticles can exchange the W boson. Therefore, the fact that C-symmetry is violated did not cause the same excitement as the discovery of P-violation.

It is important to stress that **C-symmetry violation does not mean that the properties of particles and antiparticles are different**. To justify this statement, let us consider a simple example. Suppose two companies manufacture screws for fastening metal constructions. One of them, which we call "Monolith", makes right-hand screws and labels them with the letter 'M.' Another company named "Whatever Works" (WW) produces left-hand screws and labels them with the letter 'W.' The holes in the metal constructions fastened by

M and W screws have either right-handed or left-handed threads, respectively. The Monolith company declares that the performance of its screws is the same as those of the WW company, and we want to test this claim. In the proposed example, the analogue of C-conjugation would be the replacement of a W screw by an M screw in a construction. However, M screws never fit into the holes prepared for W screws because the thread direction of the hole and the screw are different. Hence, C-symmetry related to this analogue of C-conjugation is clearly violated. But this violation does not mean that the performance of M and W screws is different. To make a proper comparison, the Monolith company should produce left-hand screws, which can be fitted instead of W screws. Only after that, we can compare their qualities. Similarly, to establish the difference in properties of particles and antiparticles, just C-conjugation alone is not sufficient. We need to apply consequently C- and P-operations. That is, we need to exchange particles and antiparticles and, after that, replace all right-handed fermions by left-handed ones and vice versa.

Let us do the specified test and apply the sequence of C- and P-conjugations to the diagram in Fig. 10.6 (left). After C-conjugation, we get the diagram in Fig. 10.6 (center). After that, P-conjugation replaces the left-handed chirality with right-handed chirality and vice versa. The obtained diagram (see Fig. 10.6 (right)) represents the interaction of the right-handed antiparticles, which satisfies all the rules of weak interaction. Thus, while the separate application of C- and P-conjugations produces the diagrams of non-existent processes, their combination results in the allowed process.

The consecutive application of C- and P-conjugation is called *CP-conjugation*. In the case of Fig. 10.6, this operation transforms the process involving particles into the process with antiparticles. What is important is that both processes are allowed by all laws of nature. In general, the same conclusion on the result of CP-conjugation can be made for all particles and all interactions. The operation of CP-conjugation always transforms left-handed particles into right-handed antiparticles, and vice versa. Consequently, a process involving particles is transformed into a process with antiparticles. The conjugated process satisfies the rules of the underlying interaction and, therefore, must exist.

Like for all other transformations, we can define *CP-symmetry* related to CP-conjugation. This symmetry can be conserved or violated. CP-symmetry is conserved if the behavior of particles and antiparticles in a particular interaction is the same. Conversely, the **violation of CP-symmetry is equivalent to the differences between particles and antiparticles**.

Here we need to make several important comments.

First, a possible violation of CP-symmetry is very different from the violation of P- and C-symmetries discussed earlier. The P-symmetry is violated because the operation of P-conjugation applied to any interaction involving W-boson exchange results in a process that cannot happen in our universe. The violation of C-symmetry is due to the same reason. However, CP-conjugation always produces an existing process. Thus, CP violation, if any, can be detected only as a quantitative difference between a process involving particles and the corresponding process with antiparticles. For example, if a meson decays more quickly than a corresponding antimeson, then CP-symmetry is violated in such a decay.

Returning to the above example of screws, we can say that P violation corresponds to the fact that the Monolith company manufactures only right-hand screws. C violation means that M screws do not fit in the constructions fastened with W screws. Finally, CP violation would be equivalent to the statement that properties of constructions fastened with left-hand M and W screws are different.

Second, although the word "violation" has a negative connotation, the violation of a symmetry always opens novel capabilities in a physical system. Hence, consequences of such violation are rather positive. In particular, if the violation of CP-symmetry indeed happens, it will result in completely new qualities of our universe. Without CP violation, all properties of particles and antiparticles must be the same, and the difference between them is possible only because of CP violation.

The violation of symmetry in our normal world also sometime creates beautiful results.

Third, the discussion presented so far does not imply that CP-symmetry must be violated in weak interactions. In principle, a universe in which both P- and C-symmetries are violated while CP-symmetry is conserved, could exist.

But the universe in which we live needs CP violation because this violation is essential for explaining the disappearance of anti-matter. All antiparticles have vanished while particles continue to exist. Thus, the properties of particles and antiparticles must differ. Without this difference, nature would not distinguish matter from antimatter, and the universe would always contain equal amounts of them both. CP violation serves as an indicator of differences between particles and antiparticles, and this role justifies the attention that particle physicists pay to this particular symmetry.

In our usual world, the violation of symmetry is straightforward to notice. For example, the missing leg of a stool represents an apparent violation of symmetry. Hence, we can predict the inevitable fall of such a broken stool without any mathematical computations, especially if someone tries to sit on it. The violation of CP-symmetry is also often easier to detect than a difference between particles and antiparticles. At the same time, these two phenomena are equivalent. That is why the efforts of scientists are concentrated on the search for manifestations of CP violation in different processes and quantitative measurement of this effect. Researchers expect that this study will help them to understand hidden forces of nature that have led to the disappearance of antimatter.

In the following chapters, several experimental measurements of CP violation are discussed in more detail. However, before proceeding to this part of the story of missing antimatter, we need to consider one more type of symmetry. It is probably the most unusual symmetry, which nonetheless plays a crucial role in particle physics. This is a symmetry related to the reversal of time.

10.3 Reversal of Time

Cinematographers often use a simple and well-known trick that consists of the reverse playback of a filmed scene. In such a playback, people on the screen move backward, and shards of ceramic, which were scattered all over the place, miraculously join together and form an unbroken vase. In science, this trick is called the *operation of time reversal*, or *T-conjugation*. The corresponding *symmetry of time reversal*, or *T-symmetry*, is conserved if the reverse video playback of a process is indistinguishable from the converse process which happens in reality. For example, a video of freezing water in reverse playback looks the same as real melting ice. Therefore, water freezing respects the symmetry of time reversal.

However, the above example is an exception rather than a rule. Most events in real life cannot be inversed without breaking T-symmetry, although people always want and persistently strive to do this. Many of us have tried at least once to glue together the pieces of a broken vase. Very often, the result of this attempt of time reversal has been very close to the original. But a closer look will inevitably reveal tiny differences between the original and restored vases, such as the cracks on the surface of the repaired vase with traces of glue on them. In contrast, the reverse playback of the vase breaking would show that all shards are collected into the original vase without any defects. This dissimilarity between the recording and the real object means that the symmetry of time reversal is broken.

In the world we live in, many processes, such as the aging of people, cannot be reversed at all. On the contrary, in the microworld of particles, all processes can be reversed in time. As always, we can justify this statement using the language of Feynman diagrams. In this language, the operation of T-conjugation replaces all particles in the initial state by the particles in the final state, and vice versa. For example, the creation of an electron-positron pair by two photons

(Fig. 5.1) after T-conjugation becomes the annihilation of an electron and positron (Fig. 5.3). Both of these processes exist. In general, the operation of T-conjugation can be applied to any diagram, and the process corresponding to the new diagram must exist.

T-symmetry in the microworld is defined similarly to the normal world. We need to imagine a reverse video playback of a certain process. If this playback is indistinguishable from the actual reverse process, T-symmetry is conserved. Otherwise, it is violated.

Like in the case of charge conjugation, T-symmetry is always conserved if the Feynman diagrams of a process and its time reversal are the same or can be transformed into one another by any rotation of fermion lines. It is apparent that by rotating the diagram in Figs. 5.1 by 180° we obtain the diagram in Fig. 5.3. Therefore, T-symmetry must be conserved in these processes. Also, after applying the operation of time reversal, the diagrams of the electromagnetic interaction of two electrons or two positrons (see Fig. 10.5) remain the same. Therefore, T-symmetry is also conserved there. In fact, the conservation of T-symmetry is a general rule for all electromagnetic interactions. The main reason for this conservation is that the vertex of electromagnetic interaction, shown in Fig. 4.2, does not change after T-conjugation. Therefore, all processes containing only these vertices are T-symmetric.

In this respect, weak interaction is substantially different. All transformations of fermions, mediated by the exchange of the W boson, can be reversed in time. For example, both the transitions $\nu_\mu \to \mu$ and $\mu \to \nu_\mu$ exist. The corresponding vertices are shown in Fig. 10.7. However, these diagrams cannot be transformed into each other by any rotation of fermion lines because the neutrino and the muon are distinct objects. Hence, T-symmetry can be violated in processes involving changes of flavor. We do not claim yet that T-symmetry is indeed violated in weak interactions, but this could just be a question of time.

Weak interaction breaks many symmetries of the world. Such behavior makes this force of nature truly unique and special. However, one type of symmetry can never be broken, even by the almighty weak force.

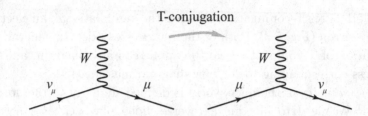

Fig. 10.7 The operation of T-conjugation applied to the transition $\nu_\mu \to \mu$.

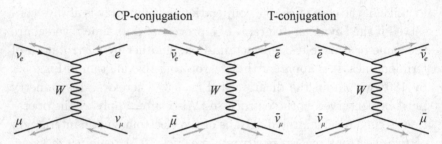

Fig. 10.8 The interaction of electron neutrino and muon (left). The process obtained after applying CP-conjugation (center). The process obtained after applying C-, P- and T-conjugations (right).

Let us consider the consecutive application of all three operations of C-, P-, and T-conjugation. The corresponding operation is called *CPT-conjugation*. For example, let us apply this operation to the interaction of a muon and an electron neutrino (see Fig. 10.8 (left)). We already established that the action of CP-conjugation results in the diagram in Fig. 10.8 (center). This diagram corresponds to the interaction of an antimuon and an electron antineutrino. The processes in Fig. 10.8 (left) and Fig. 10.8 (center) can be different if the properties of particles and antiparticles are not the same. If present, this dissimilarity corresponds to CP violation. Let us now perform the operation of T-conjugation on the diagram in Fig. 10.8 (center). The action of this operation results in the replacement of all particles in the initial state by the particles in the final state and vice versa. Consequently, we obtain the diagram in Fig. 10.8 (right). It corresponds to the interaction of a muon antineutrino and positron.

On the other hand, the rotation of the diagram in Fig. 10.8 (left) by 180° results in the diagram shown in Fig. 10.9 (right). The

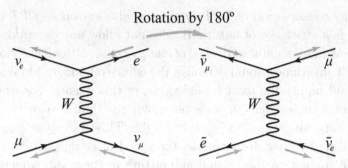

Fig. 10.9 The interaction of electron neutrino and muon (left). The rotation of this diagram by 180°.

only difference between the rotated diagram and that in Fig. 10.8 (right) is the vertical order of the fermion lines.[3] However, this difference is inessential because the vertical order of lines never influences depicted processes. Hence, the rotation by 180° effectively transforms the diagrams in Fig. 10.8 (left) and 10.8 (right) into one another. The possibility of such a transformation implies that the symmetry corresponding to CPT-conjugation (that is, *CPT-symmetry*) must be conserved. This conclusion is not changed even if CP- and T-symmetries, are separately violated.

A similar consideration of CPT-conjugation can be done for any process. The result will always be the same. This conjugation always transforms an initial diagram into the new one, such that one of them can be obtained from another by some rotation of fermion lines. This means that CPT-symmetry is conserved in all processes and all interactions. Even though the C-, P-, and T-symmetries can be individually violated, the consecutive application of C-, P-, and T- operations always respects CPT-symmetry. This result is known as the *CPT theorem*. It constitutes one of the most important statements of particle physics and can be rigorously proven using mathematical methods. However, the language of Feynman diagrams presents an easy and illustrative method to verify the validity of the CPT theorem without any involvement of mathematics.

---

[3]In Fig. 10.8 (right), the line corresponding to the transition $\nu_e \to e$ is above the line corresponding to $\mu \to \nu_\mu$, and in Fig. 10.9 it is below.

The conservation of CPT-symmetry, also known as *CPT invariance*, is a strict law of nature. It does not allow any exception. This law is as fundamental as the law of energy conservation. The violation of CPT invariance would demolish the edifice of modern physics since it would nullify the most basic axioms of this science. For example, physics assumes that an observer moving with any velocity should always measure the same speed of light. This statement may seem surprising because, for example, the speed of a train is different for people staying on the ground and sitting in the train. Nonetheless, the constant value of the speed of light has been confirmed by numerous experiments and is considered to be one of the most fundamental properties of nature. If CPT-symmetry were violated, this law would also be broken. That is why the possibility of CPT violation is thought to be impossible. Experimenters, as always, do not take the CPT theorem at face value and try to find possible deviations from this law. But all such attempts have not succeeded yet.

There is one important consequence of the CPT theorem. CPT invariance means that if CP-symmetry is violated in a certain interaction, then T-symmetry must also be violated there. If CP-symmetry were violated while T-symmetry were conserved, the consecutive application of CP and T operations would result in CPT violation. That is why CP and T violations are, in fact, equivalent. This equivalence may look unexpected since the action of the two operations is entirely different. CP violation equates to different properties of particles and antiparticles. For example, the decays of neutral kaon and anti-kaon to two pions, presented in Fig. 9.6, can differ. The violation of T-symmetry means the distinction between direct and reverse transitions of fermions, such as that shown in Fig. 10.7. These phenomena seem not to be connected at all. However, because of the CPT theorem, CP and T violations are intrinsically related, so that both are equally crucial for understanding the disappearance of antimatter.

From an experimental point of view, verification of CP-symmetry is much easier to do. CP violation was discovered more than 50 years ago. The violation of T-symmetry in the microworld has also been proved, but it happened only in 2012. Thus, it is not surprising that differences between particles and antiparticles have mainly been studied using CP violation. Still, we will demonstrate in Section 15.1 that the origin of CP violation is easy to understand using its connection to T violation. That is why both these symmetries are important for our story.

Concluding this chapter, we can summarise the results obtained. We considered three important operations: P-, C- and T-conjugations. Symmetry with respect to these operations or its violation is essential for understanding the behaviour of particles. Weak interaction violates both P- and C-symmetry, because only left-handed particles and right-handed antiparticles participate in the interactions with W boson. However, violation of CP symmetry is more difficult to establish. If proven, this violation will imply differences in properties of particles and antiparticles. Because of the CPT theorem, CP and T symmetries are equivalent.

Now that the relation between different types of symmetry is established, we can consider the discovery of CP violation in more detail. This discovery, undoubtedly, has become the first important step in understanding the mystery of the vanished antimatter. Therefore, it deserves special attention.

Chapter 11

Discovery of CP Violation

The experimental confirmation of CP violation was obtained in 1964. At that time, most scientists did not doubt the conservation of CP-symmetry. This belief was based on a firm conviction that the behavior of particles and antiparticles must be the same.[1] Certainly, this stood in contradiction to the absence of antimatter in the universe, but nobody realized this then.

Throughout the history of science, there have been many examples of an experimental test of a seemingly undeniable truth resulting in a new discovery, which has ultimately changed the views of people on this verity in particular, and on nature in general. The spirit of nonconformity always drives science forward. Scientists challenging canonical concepts and trying to disprove them always emerge amidst a bulk of scholars adhering to a commonly accepted doctrine. Of course, at most, only one out of many thousand attempts to overturn established rules leads to victory. But the rarity and presumed impossibility of success never stops challengers. The obstacles just amplify the triumph of the winners whenever a positive result is achieved. The unearthing of the difference between matter and antimatter represents one of many such accomplishments, which have refuted adopted principles and modified the course of physics.

---

[1]As was explained in the previous section, equal properties of particles and antiparticles imply the conservation of CP-symmetry.

The experiment which resulted in the discovery of CP violation was performed by a group of scientists led by the American physicists James Cronin and Val Fitch. Before proceeding with the description of their investigation, we need to consider in detail the particle which they studied. These details are quite technical, but they are necessary for grasping the idea behind Cronin and Fitch's experiment and understanding the obtained results.

In their research, Cronin and Fitch explored the decays of a neutral kaon into two pions. The corresponding diagram of this decay is shown in Fig. 9.6. At first glance, it would seem that a neutral kaon would not be a suitable particle to use when searching for differences between matter and antimatter. Indeed, both the neutral kaon and antikaon decay to the same final state, that is, to a pair with one positive and one negative pion. Therefore, the detection of the decay products by itself says nothing about the particle or antiparticle nature of their parent.

However, a neutral kaon has a very unusual property, which was instrumental in the discovery of CP violation. A neutral kaon can transmute into its antiparticle (neutral antikaon). This statement sounds very strange because, in all processes considered so far, a particle and antiparticle can annihilate but never turn into one another. Indeed, this phenomenon puzzled scientists for a very long time, and understanding it in full became possible only after the advent of the quark model. The transformation of an electron into a positron is impossible because it would violate the law of charge conservation.[2] However, the charge is not the only barrier that prevents the conversion of a particle into an antiparticle in most cases. A neutron is a neutral particle, and, still, it cannot turn into an antineutron either.[3] The conversion of a particle into its antiparticle is possible only for some neutral mesons, such as the neutral kaon.

---

[2]The charges of an electron and positron are opposite.

[3]In fact, the claim that a neutron cannot transform into an antineutron is stated by the Standard Model. As usual, some scientists try to detect this transformation and disprove the Standard Model. These attempts will be discussed in the following chapters.

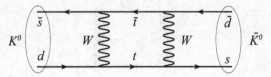

Fig. 11.1   Transformation of a neutral kaon (K^0) into its own antiparticle ($\bar{K}^0$).

The transition of a kaon into an antikaon is described by the diagram in Fig. 11.1. This diagram indicates that after a chain of flavor transformations mediated by W bosons, the initial state $d\bar{s}$ forming a neutral kaon changes into the final state $s\bar{d}$, which corresponds to a neutral antikaon. The lower fermion line represents the conversion of the down quark into a top quark (transition $d \rightarrow t$) followed by the transition of the top quark into a strange quark ($t \rightarrow s$). The upper fermion line is subject to the same chain of transitions ($d \rightarrow t$ and $t \rightarrow s$). It develops back in time because it corresponds to antiquarks.

The process presented in Fig. 11.1 is quite unusual because both the initial and final states contain just one meson, and no other particles are emitted or absorbed. Such interactions are possible only if the masses of the initial and final particles are the same. Otherwise, the energy would not be conserved. The mass of any particle and its corresponding antiparticle are always equal. Hence, the requirement of energy conservation is indeed satisfied in such a process.

The interaction shown in Fig. 11.1 means that the neutral kaon can become its antiparticle sometime after production. Later, the antikaon can, again, turn into a kaon due to the similar quark transformations shown in Fig. 11.2. This property of neutral kaons is called *oscillation*. The kaon, like a pendulum, oscillates between the states of particle and antiparticle and these transformations continue until its decay. The oscillation is a *random process*, like all other processes in the microworld of particles. In other words, at any given moment in the neutral kaon's life, only the *probability* for it to be a particle or antiparticle is known.

The concept of probability is widely used in modern life. When a weather forecast says that the chance of rain is 80%, we know that an umbrella will be needed for a walk outside in four cases out of five. The meaning of the phrase "the probability is 60% for a neutral kaon

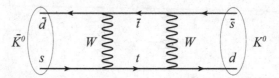

Fig. 11.2   Transformation of a neutral antikaon ($\bar{K}^0$) into its own antiparticle (K^0).

to be a particle" is similar. If we determine the quark content of many neutral kaons with the same properties, we will find this content to be $d\bar{s}$ in three cases out of five. In the other two cases, it will be $\bar{d}s$. This probability varies during the kaon's lifetime. Sometimes, the probability for it to be a particle increases to 100%; sometimes, it drops to zero, but, in general, there is always a chance for an object called a neutral kaon to be identified as either a particle or antiparticle. Because of the continuous transitions between the states of particle and antiparticle, the neutral kaon and antikaon *mix* and form two new objects.

In our everyday world, a mixture of several materials produces a new material with distinctive properties. These properties may be significantly different from that of the original substances. For example, bronze is an alloy of copper and tin. It is more durable than copper, and this property was key in the development of ancient civilizations that had discovered bronze and started to use it many thousand years ago. The mixtures of several particles can also form other objects, which possess new qualities. A meson can be treated as a mixture of a quark and antiquark, and the behavior of a meson is entirely different from that of its constituents. Some mesons represent an even more complicated mixture. For example, a *neutral pion*, or π^0 *meson* is a mixture of $u\bar{u}$ and $d\bar{d}$ components.

The mixing of a neutral kaon and antikaon is unlike the combination of a quark and antiquark, which produces a meson. Any quark and antiquark can form a meson[4] because of the attractive strong force acting between them. In its turn, the mixing of a meson and

---

[4]except a top quark, which decays before any meson containing it can be created.

antimeson is possible only if they oscillate, that is, transform into each other. For example, positive and negative pions never mix and exist as separate particles because they do not oscillate. In fact, just three other mesons, in addition to the neutral kaon, can oscillate and, consequently, mix. We will discuss some of them later.

Two new particles produced by the mixing of a neutral kaon and antikaon are called the *short-living* and *long-living kaons*, or K_S and K_L mesons, respectively. These names suggest that the lifetimes of K_S and K_L mesons are considerably different. This is indeed the case. The distinction between them is caused by the dissimilarity of the K_S and K_L *decay modes*, that is, the final states of their decays. More precisely, the difference is produced by the specific properties of K_S and K_L mesons related to the operation of CP-conjugation.

Both K^0 and $\bar{K}^0$ mesons can decay to a pair of oppositely charged pions (see Fig. 9.6). However, it turns out that if CP-symmetry is conserved, only the K_S meson will decay to this final state. For the K_L meson, the decay to π^+ and π^- will be forbidden. Conversely, the main decay mode of the K_L meson should be the group of π^+, π^-, and π^0 mesons.

In the above statement, the condition of CP conservation is crucial because, if CP-symmetry is violated, the K_L meson will be able to decay to two pions. Likewise, the observation of the decay of the K_L meson to two pions will signify the violation of CP symmetry. Here comes the idea by Cronin and Fitch. They aimed to discover CP violation by detecting the decay of K_L meson to a pair of positive and negative pions. To convey the intent of Cronin and Fitch, we needed to introduce the complicated concepts of mixing of neutral kaons which produces K_S and K_L mesons. But as far as this effect becomes known, the plan for the discovery of CP violation turns out to be transparent.

The experiment of Cronin and Fitch was carried out at a proton accelerator at Brookhaven National Laboratory in the USA. Accelerated protons collided with a target and created many new hadrons. Among them, there were neutral kaons and antikaons. Because of their oscillation, these particles mixed and produced a beam of K_S and K_L mesons.

Both K_S and K_L mesons live, on average, for less than a microsecond. However, if they move with high speed, they can travel large distances. The mean lifetime of the K_S meson is about 500 times less than that of the K_L meson. The shorter lifetime means that the average flight distance of a K_S meson is much smaller. This distinction between the K_S and K_L mesons was used in Cronin and Fitch's experiment. Their apparatus, which detected decays of neutral kaons, was placed at about 17 meters from the point of kaon production. Only K_L mesons survived at such a distance, while all K_S mesons decayed much earlier.

If CP-symmetry were conserved, the decay of the K_L meson to two oppositely charged pions would be prohibited. Indeed, the main K_L decay mode detected in the experiment was a group of π^+, π^- and π^0 mesons. This decay is allowed by CP conservation. But rarely, the researchers also recorded the decays of a K_L meson into π^+ and π^- mesons without π^0. Such decays happened in about one case out of five hundred. Irrespective of the low probability, simply being able to observe such a process signified that CP-symmetry in K_L decays was violated. More precisely, CP-symmetry was violated in weak interaction, which governs the decay of K_L meson.

We established in the previous chapter that CP violation is equivalent to differences between particles and antiparticles. On the other hand, the detection of CP violations turned out to be much easier than finding the dissimilarity between properties of particles and antiparticles. To uncover CP violation, just the observation of the specific decay mode of the meson is sufficient. However, the measurement of differences between particles and antiparticles requires much more effort. It is sufficient to say that the first detection of such a difference in the decays of mesons and antimesons was reported about forty years after the discovery of Cronin and Fitch. This fact highlights the significance of CP violation for explaining the disappearance of antimatter and the importance of Cronin and Fitch's result. As can be expected, these two scientists were rewarded with the Nobel Prize in 1980 for their achievement.

We need to stress that the proof of CP violation by itself does not clarify the reasons for the absence of antimatter. Above all, the CP violation observed by Cronin and Fitch is a small effect. The bulk of K_L decays honor CP-symmetry, and only one decay in 500 defies it. In this respect, CP violation is entirely different from the violation of P-symmetry: 100% of decaying muons are left-handed so that 100% of muon decays violate the P-symmetry. Thus, CP violation, considered alone, cannot explain a huge imbalance between matter and antimatter. It seems that this violation is just one of the necessary conditions for such an explanation. In other words, it would be impossible to justify the excess of matter in the universe without CP violation. But CP violation is not the only requirement. Particles must possess several other properties, which are essential for explaining the disappearance of antimatter.

Nonetheless, the discovery of CP violation is a giant step forward. Before Cornin and Fitch's experiment, scientists did not consider the problem of missing antimatter at all. Active research in this direction began only after the differences between particles and antiparticles were experimentally proven. In particular, the detection of CP violation has motivated the formulation of prerequisites for the dominance of matter in the universe. These requirements, known as Sakharov's conditions for baryogenesis, are essential for our quest because they define the research direction to follow. All of them are considered in the next chapter.

Chapter 12

Sakharov's Conditions

12.1 Violation of Baryon Number

Andrei Sakharov lived in the now non-existent country, the Soviet Union. He is famous because of his many accomplishments. Sakharov was one of the developers of the Soviet nuclear weapon. He participated in the creation of the most powerful explosive device in the human history called the "Tsar-bomb." The destructive energy released by this weapon exceeded that of the atomic bomb dropped on Hiroshima, by more than 3000 times. Sakharov also devoted a considerable part of his life to defending human rights. He was persecuted by the Soviet government for this, spent several years in exile, and subsequently was awarded the Nobel Peace Prize for his efforts in this domain. But for our story, another of Sakharov's achievements is important. He was the first to realize the significance of CP violation in the disappearance of antimatter. And he went even further. Shortly after the discovery by Cronin and Fitch, Sakharov formulated three requirements necessary for baryogenesis, that is, for the dominance of matter in the universe. This work of Sakharov's, which is now considered as one of his main results, remained unnoticed for a long time. Only after several dozen years, the role of *Sakharov's conditions* in unraveling baryogenesis was recognized and praised by the scientific community. Currently, effectively all papers on baryogenesis mention or refer to these conditions. This example confirms once again that ideas don't vanish, just as manuscripts don't burn.[1]

Let us consider Sakharov's conditions in detail and try to understand why their fulfillment is obligatory for explaining the disappearance of antimatter. For this, we need to recall the main properties of all interactions. All transformations of particles, including annihilation and creation, are governed by interactions. That is why interactions are the keys to unlocking the chest that hides all the mysteries of nature.

In the previous chapters, we discussed three different types of interactions: electromagnetic, weak and strong. In addition to them, there are also Higgs and gravitational interactions. The strength of both of them depends on the mass of participating objects. However, the basic theoretical principle states that the masses of a particle and a corresponding antiparticle must be equal. Therefore, the Higgs interaction should not distinguish matter from antimatter. Gravitational properties of particles and antiparticles are also expected to be the same. To be fair, this expectation has not been experimentally verified so far. Currently, several experiments with antiparticles are being carried out at CERN (see Section 3.4 for details). They aim to test the gravitational interaction of antimatter with matter. The observation of any anomaly in this interaction would be one of the most fundamental discoveries, which would completely change our understanding of nature. However, presently not many scientists expect such a result. Consequently, discoveries regarding the gravitational force that would help explain the disappearance of antimatter are not anticipated.

Let us consider the properties of other known interactions. The electromagnetic and strong forces are very different. Still, they have several similar features. In particular, both never change the flavor of particles. An electron always remains an electron after electromagnetic interaction, and quarks keep the same flavor after exchanging gluons. In this respect, the weak interaction is singled out, because

---

[1]The phrase "Manuscripts don't burn" comes from the novel *The Master and Margarita* by the Russian writer Mikhail Bulgakov. This book was published only after his death.

it changes the type of fermions. An up quark can become a down quark, and an electron can turn into a neutrino. All fermions can change their flavor by participating in weak interactions.

Regardless of these differences, all known interactions have an important common characteristic. A quark after an interaction continues to be a quark. It cannot disappear by itself or become a lepton. This feature is reflected in the empirical rule of pairs, which was discussed in Section 6.1. According to this rule, a quark always starts its existence in a pair with an antiquark. Also, the destruction of a quark is possible only together with an appropriate antiquark. A similar statement is also valid for leptons. A lone lepton can never appear or disappear. Consequently, **the known interactions cannot change the difference between the total number of quarks and antiquarks in any physical system.** Quarks and antiquarks can unite into hadrons, and hadrons can either decay or exist eternally. However, the difference between the number of quarks and antiquarks never changes in these processes. For example, in the decay of the charged pion, shown in Fig. 9.5, there is one quark and one antiquark in the initial state, and there are no quarks in the final state. Instead, a pair involving a lepton and an antilepton appears. Thus, the difference between the number of quarks and antiquarks remains the same before and after the weak interaction. The same applies to the difference between the number of leptons and antileptons.

The universe, considered as a whole, also represents a physical system. Consequently, none of the known interactions can change the balance between quarks and antiquarks in the universe during the entire time of its existence. An equal number of quarks and antiquarks must have been created after the Big Bang, and the same equality should remain today. This conclusion is contradictory to experimental results, which do not show any significant amount of antimatter anywhere in the observable universe. The experiments prove that the number of atoms containing quarks and electrons is much higher than the number of antiatoms made of antiquarks and positrons. Moreover, antiatoms are not detected in space at all (see Section 6.2 for details). On the other hand, the properties of quarks and antiquarks are essentially the same, except for a tiny difference between them reflected by CP violation. This violation of CP symmetry is indeed minute. In any case, it cannot allow antiquarks to exist as free particles[2] and cannot prevent antiquarks and positrons from creating antiatoms. Thus, an experimentally established significant imbalance between atoms and antiatoms means that the number of antiquarks existing in the observable universe is much lower than the number of quarks.

There is only one possible resolution to the apparent paradox: the rule of pairs must be violated. **There must be some special type of interaction that causes the creation of one or several quarks without the corresponding antiquarks.** Only such an interaction can make a difference between the total number of quarks and antiquarks and, consequently, give rise to a dominance of matter in the universe. This requirement constitutes *Sakharov's first condition* for baryogenesis.

The rule of pairs is an empirical law. It may look obvious and natural, but still, it is not a fundamental law of nature. It does not follow from any core principles, which form a foundation for our understanding of nature. The rule of pairs summarizes all experimental results obtained so far. But not a single experiment can guarantee that this rule will be respected in all future measurements. Moreover, according to Sakharov's first condition, there must exist processes violating the rule of pairs.

Sakharov did not use the concept of quarks for defining his first condition. His work was published in 1967. At that time, the quark theory had already been formulated by Gell-Mann and Zweig but had not yet been proven. Consequently, it was treated by many scientists as one of the promising theoretical models but not as an established theory. It is sufficient to say that Gell-Mann himself initially considered quarks as mathematical constructions, which helped to explain the structure of mesons and baryons but did not correspond to real particles. That is why Sakharov formulated his first condition in terms of a *baryon number*. The definition of this number is simple. Each baryon, such as a proton or neutron, is assigned the value of

---

[2]They must form antihadrons the same way quarks unite into hadrons.

+1, while each antibaryon is assigned −1. The baryon number of each meson is zero. The baryon number of all leptons is also zero. For any group of particles, the total baryon number is the arithmetic sum of the individual values and is equal to the difference between the number of baryons and antibaryons. Since the baryon number of all mesons and leptons is zero, the presence of any number of these particles does not change the total baryon number.

Each baryon contains three quarks, and each antibaryon is made of three antiquarks. Therefore, the non-zero baryon number is equivalent to the non-zero difference between the number of quarks and antiquarks. Mesons cannot influence this difference because all of them comprise a quark and antiquark. The *conservation of baryon number* in a particular process means that the difference between quarks and antiquarks remains constant during the process, and the *violation of baryon number* indicates a variation of this difference.

The baryon number is conserved in all known interactions. In particular, each created baryon is always accompanied by the corresponding antibaryon, and the destruction of a baryon always happens together with an antibaryon. It is easy to check the conservation of baryon number in all processes discussed so far. However, **Sakharov's first condition demands the violation of baryon number**. Particles must have some special properties and participate in unusual interactions, which produce quarks without antiquarks.

Thus, Sakharov's first condition is not just a collation of known properties of nature. This condition predicts the existence of entirely new physical phenomena, which hide beyond the territory explored. What is even more amazing, the prediction of new physics is made by using simple logical reasoning and without involving complicated mathematical formulae.

Knowing that something new and unknown must exist is very exciting. This knowledge supports scientific research and motivates the development of science. That is why the consequences of Sakharov's first condition go far beyond the problem of antimatter.

It is important to stress that the first condition is not based on an assumption of an equal number of primordial quarks and antiquarks created after the Big Bang. Supposing that there had been an excess of primordial quarks, this initial imbalance must have been generated by some interaction. Even if the corresponding processes happened during the Big Bang, they still must have violated the baryon number. Thus, the first condition can never be circumvented. Its fulfillment is mandatory for creating the excess of matter in the universe.

Leptons can also be attributed a *lepton number* similar to the way the baryon number is defined for hadrons. Each lepton, both charged (e.g., electron or muon) and neutral (that is, neutrino), is assigned the value of +1, and each antilepton is assigned −1. The lepton number of a group of particles is equal to the difference between the number of leptons and antileptons. The lepton number is conserved in all known interactions. For example, in β decay shown in Fig. 9.4, an electron is produced together with an antineutrino. Consequently, the lepton number before and after decay is equal to zero. Conversely, the production of an electron and neutrino in β decay would mean a change of lepton number by 2. However, such a decay has never been observed.

Sakharov's first condition requires the violation of baryon number and says nothing about the lepton number. This is not accidental. On the contrary, there are serious reasons for this. Each atom contains both baryons and electrons. Therefore, the dominance of matter in the universe also signifies an excess of electrons over positrons. But this excess is not automatically equal to the difference between leptons and antileptons. The group of leptons includes invisible neutrinos, and, currently, information on the proportions of neutrinos and antineutrinos across the universe is insufficient.

The conclusion on the absence of antiatoms in space is based on our ability to detect the annihilation of antiparticles — see Section 6.2 for details. However, for neutrinos, this argument does not work. Neutrinos are very special particles. They are sensitive only to weak forces. The annihilation of a neutrino and antineutrino never produces photons, because neutrinos do not interact electromagnetically. The only possible mode of neutrino-antineutrino annihilation would be for them to turn into another fermion-antifermion pair. But in some cases, even this annihilation is forbidden. After neutrinos, the lightest fermion is an electron. If the total energy of the interacting neutrino and antineutrino is below the rest energy of an electron and positron, then neutrino annihilation is not possible at all because of the law of energy conservation. This veto means that, in principle, space can be filled with low-energetic neutrinos and antineutrinos, which peacefully coexist and do not destroy one another. Such harmony is impossible for all other fermions, since they are charged particles and, therefore, always have the possibility of annihilating into two photons.

The low-energetic neutrinos rarely interact with other particles. Even the action of gravitational force on neutrinos is extremely gentle because of the small neutrino mass. Therefore, the measurement of the relative proportions of neutrinos and antineutrinos in space is currently too imprecise. This handicap leaves open the possibility of an excess of antineutrinos over neutrinos across the universe. In the case of such an excess, the deficit of positrons compared to electrons would be compensated by a larger number of antineutrinos, so that the total lepton number within the universe would be zero. In other words, modern science does not have any proof for lepton number violation, neither in the microworld nor on cosmic scales. On the contrary, the violation of baryon number is confirmed experimentally by a considerable excess of atoms over antiatoms. That is why Sakharov's first condition requires baryon number violation and does not put any restriction on the lepton number.

12.2 Difference between Particles and Antiparticles

The violation of baryon number is not the only condition needed for baryogenesis. If the properties of particles and antiparticles are the same, then for each process that creates new baryons, there must be a similar process that produces new antibaryons. Let us consider a simple example. Suppose, a hypothetical interaction allows an antiproton to decay to an electron and a meson. Such decay violates the baryon number because this number changes from -1 in the initial state to zero in the final state. Thus, the first condition is satisfied, and we could suppose that this interaction would eliminate all antibaryons. However, if the properties of particles and antiparticles were the same, a proton would also decay in a similar way, producing a positron and antimeson. Moreover, the rates of proton and antiproton decay would be the same. This equality means that the average number of protons and antiprotons that have decayed during a sufficiently extended period would be equal. Consequently, no net difference between quarks and antiquarks would be generated. This example justifies *Sakharov's second condition* for baryogenesis, which states that **properties of particles and antiparticles must be different**. In other words, to destroy one group of objects and keep the other, nature must have a way of distinguishing between them.

The task of separating similar objects into several groups with entirely different destinies emerged not only during the evolution of the universe. Humankind has attempted to solve a similar problem many times during its multi-millennial history. The "objects" in these attempts were humans themselves. In this vein, it is interesting to mention one solution to this problem proposed by a certain abbot Arnaud Amalric, who lived in the thirteenth century. The name of Amalric has remained in human history because of his participation in the infamous Albigensian Crusade against a heretical Christian sect, which was established in the South of France. When Crusaders had seized the city of Béziers and had started to massacre more than 20000 of its inhabitants, Amalric was asked, by one Crusader, how righteous men should be distinguished from heretics. Amalric suggested to kill them all and let God sort them out. Obviously, the separation of people into two groups required some distinctive feature that would allow for differentiating between them. Of course, such a distinction between people of different religious faiths does not exist. Therefore, the zealous cleric preferred to shift the insurmountable burden of separation onto the shoulders of God.

This method of Amalric's can also be tried for distinguishing matter and antimatter. Nobody denies that divine power is capable of doing everything. Certainly, it could selectively destroy all antimatter even without any distinction between particles and antiparticles. However, science, contrary to religion, always aims to relieve God from dealing with problems that can be sorted out without his involvement. In the case of separating particles and antiparticles, this is possible only if their properties are different.

It may seem that particles and antiparticles are easy to distinguish. For example, they have opposite charges. An electron is charged negatively, while a positron is positive. However, the behavior of particles with positive and negative charges is the same in all electromagnetic interactions. In particular, this means that the repulsive force of two positive particles is not distinguishable from that of two negative particles. Therefore, it is impossible to find out the charge of particles by observing their repulsion. Mathematically, this means that C-symmetry, that is, the symmetry of charge conjugation, is conserved in electromagnetic interactions. In a universe with only the electromagnetic force acting between particles, it would be impossible to distinguish particles from antiparticles. An external observer studying this imaginary world would establish that particles and antiparticles had opposite charges but would not find any criteria to prefer one group or another. They would be entirely equivalent. Thus, it would be impossible to find a reason for destroying just one of them.

We established in Chapter 10 that the difference between particles and antiparticles is equivalent to CP violation. Therefore, Sakharov's second condition requires the violation of CP symmetry. To be more precise, this condition requires the violation of both CP- and C-symmetries. We could imagine a hypothetical situation, when CP violation is caused by the violation of P-symmetry, while C-symmetry is conserved. In this case, the particles and antiparticles do not differ. Hence, the exact formulation of the second condition stipulates that both C and CP symmetries must be violated. However, in our usual world, C violation is granted for free by the left-handed nature of weak interaction, so the main emphasis of Sakharov's second condition is really CP violation.

Like the first condition, the requirement of CP violation can be derived using only logical reasoning. No complicated mathematical formulae are needed. This observation means that the prediction of CP violation, which is undoubtedly one of the fascinating phenomena of nature, could have been made long before the experimental discovery of this effect by Cronin and Fitch. In reality, this has not happened. On the contrary, it was the experimental discovery of CP violation that stimulated Sakharov to explore the conditions for baryogenesis. Nonetheless, this does not diminish the importance of Sakharov's work which established the relation between baryogenesis and CP violation. Consequently, the work of Sakharov paved a way to understanding the dominance of matter in the universe.

The role of CP violation in explaining the absence of antimatter justifies particle physicists' huge interest in this phenomenon and its thorough study. The main goal of this study is to reveal the origin of CP violation. This effect has been observed in the decays of neutral kaons. However, other occurrences of CP violation in the decays and properties of other particles must also exist. The exploration of these manifestations represents one of the most attractive directions of research in particle physics. As usual, scientists want to collect all possible information on CP violation and build a comprehensive theory of this phenomenon. Certainly, one of the outcomes of this research may be the discovery of the reasons for the disappearance of antimatter. But many other results, which cannot even be dreamt of today, are not excluded.

12.3   Departure from Thermal Equilibrium

The violation of both CP symmetry and baryon number is still not sufficient to generate the imbalance between particles and antiparticles. Let us consider one more example. Suppose that there is an interaction mediated by a hypothetical boson Q and that this boson has mass and can decay, producing either a proton and electron or an antiproton and positron. The corresponding diagrams[3] are shown in the upper part of Figs. 12.1. Also, suppose that Q-boson decays generate more particles than antiparticles.

---

[3]Note the unusual direction of arrows on the fermion lines in these diagrams. This singularity once more underlines the facts that known interactions cannot conduct such processes.

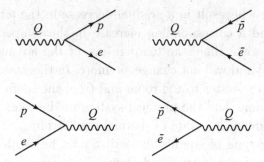

Fig. 12.1   Decays of a hypothetical Q boson to a proton (denoted as p) and electron or to an antiproton and positron (upper two diagrams). Reverse processes of creation of the Q boson from a proton and electron or from an antiproton and positron (two lower diagrams).

In the proposed model, Sakharov's first and second conditions are satisfied. Indeed, in the decays of the Q boson, the initial state contains no quarks, and therefore its baryon number is equal to zero. The final state contains either a proton or antiproton and its baryon number is either $+1$ or -1. Also, the excess of particles in Q decays signifies CP violation. However, this model still does not produce an imbalance between particles and antiparticles in the universe. The reason for this is the following.

Each process in the microworld can be reversed. So, there must exist reverse processes, in which a proton and electron, or an antiproton and positron, annihilate and produce the Q boson as shown in the lower two diagrams of Fig. 12.1. The combination of all these interactions would result in a universe containing both particles, antiparticles and Q bosons, which would continuously transform into one another. According to general laws of physics, this mixture must finally attain a state of *thermal equilibrium*. In this state, the fractions of particles and antiparticles must be constant and equal to each other.

The concept of thermal equilibrium is well-known to all of us, even though we do not always name it this way. Let us suppose you bring a glass of boiling water in a room. The air in the room and the water in the glass have different temperatures. Therefore, the excess thermal energy contained in the water will be transferred to the air.

This transfer will result in a gradual decrease in the temperature of the water and a corresponding increase in the temperature of the room. After some time, the temperature of the air and water will become equal and will not change anymore. In this state, the flow of energy from the water to the room and from the room to the water will be the same, and the physical system of the room and glass of water will remain in a state of thermal equilibrium.

A similar type of equilibrium will finally be established in the model of the universe described above. To understand why it happens, let us consider an extreme case. Supposing, a certain part of space in the hypothetical universe contains only particles (protons and electrons) and no antiparticles at all. The protons and electrons will interact with each other and produce the Q bosons as per the diagram in Fig. 12.1 (lower left). These bosons will decay either to particles (diagram in Fig. 12.1 (upper left)), or antiparticles (diagram 12.1 (upper right)). Additionally, there will be processes, in which two protons are converted into two positrons, or two electrons become two antiprotons (see the upper diagrams in Fig. 12.2). Consequently, antiparticles will appear in this part of space, while the number of particles will decrease. These transitions will happen

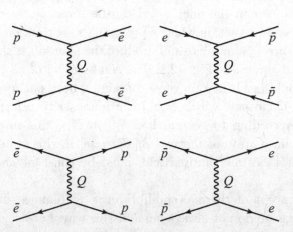

Fig. 12.2   Conversion of particles into antiparticles in a hypothetical model with Q boson (upper diagrams). Reverse conversion of antiparticles into particles (lower diagrams).

continuously. Antiprotons and positrons will emerge instead of protons and electrons. The antiparticles will also interact and produce either particles or antiparticles. As long as the fraction of particles is higher than the fraction of antiparticles, the number of antiparticles will gradually increase, while the number of particles will simultaneously decrease. These transitions will end only when the fractions of particles and antiparticles become equal, and the processes of creation of particles and antiparticles balance each other. In other words, the physical system will arrive at the state of thermal equilibrium. The main reason for this equilibrium is the possibility of *direct* (such as the decays of the Q boson) and *reverse* (such as the creation of the Q boson) processes.

The state of thermal equilibrium is not just a feature of the above hypothetical model. In all cases, when direct processes create more particles than antiparticles, there must be reverse processes which produce an opposite effect and eliminate the excess generated. The net result of all these interactions will be thermal equilibrium, in which the number of particles and antiparticles will be equal.[4]

Here we need to stress two important points.

First, thermal equilibrium is always attained whenever direct and reverse processes are possible. This is a general law of physics, which works in any physical system and not only in the microworld of particles.

Second, the state of thermal equilibrium does not happen instantly. To reach it requires a substantial duration of time. Also, the number of direct and reverse processes, which occur in the physical system, should be sufficiently high. In the above example of hot water in a room, the temperatures of the room and water do not immediately become equal. Similarly, some time is needed to equalize any physical system.

---

[4]In general, thermal equilibrium only implies that the fractions of particles and antiparticles in the physical system are constant. The key reason for equality of these fractions is the fact that the mean energy of particles and antiparticles is the same. In turn, the equality of energies is explained by equal masses of any particle and the corresponding antiparticle.

In any case, **if both direct and reverse processes are possible, thermal equilibrium is inevitable**. Hence, the proportions of particles and antiparticles will always be equalized after some time, even if the first and second conditions generate an initial excess of one of these fractions. Therefore, *Sakharov's third condition* for baryogenesis states that **processes that produce an excess of particles over antiparticles must be out of thermal equilibrium**. Essentially, this requirement means that at a certain stage in the universe's evolution, only the processes of particle creation happened, while the reverse processes of particle annihilation were suppressed.

That's it. The three Sakharov's conditions have been formulated. The fulfilment of all of them is essential for the generation of the excess of matter in the universe. In the next part of this story, we will consider several theoretical models, which satisfy these conditions and hence propose realistic schemes of baryogenesis. But before advancing to the third part, let us draw some intermediate conclusions.

We have come a long way in trying to resolve the problem of disappeared antimatter. We have started by defining the question, which needs to be answered. We have gradually collected knowledge about particles and their properties. We have considered three types of interactions and have realized their role in transforming the primordial mixture of quarks and leptons into atoms, which compose all known materials. Using the main properties of these interactions and applying the power of logical reasoning, we have obtained three main conditions necessary for generating the predominance of matter in the universe. We should continue to move in this direction. As stated above, there are realistic theoretical models that satisfy all three of Sakharov's conditions and can, therefore, explain the disappearance of antimatter. So, we need to consider these models and the status of their experimental verification.

We can already say now that none of the models provide the final answer. Thus, explaining the imbalance between matter and antimatter still requires some work to be done. We need to continue the study of particles and their properties. However, there is a reward at the end of the road. The results of the exploration of particles should exceed the most ambitious expectations. They will reveal the universe's deepest mysteries and uncover new fascinating phenomena, which presently seem unimaginable and even completely impossible. This outcome is guaranteed by the fact that the known properties of particles cannot satisfy Sakharov's conditions for baryogenesis. Therefore, new discoveries are forthcoming; we can count on this.

Now, it is time to turn this page and proceed to the next part of the book.

PART 3

The Genesis of Matter

"There are more things in heaven and earth, Horatio,
Than are dreamt of in your philosophy."
William Shakespeare, *Hamlet*

Chapter 13

Baryons and Photons

The main outcome of the previous discussion consists of defining the essential conditions, which are necessary for the dominance of matter in the universe. These requirements were first formulated by Sakharov. Therefore, they rightfully bear his name. With these conditions satisfied, the origin of the imbalance between matter and antimatter becomes clearer and more understandable. Let us suppose there is a special interaction, which creates quarks without antiquarks or vice versa (the first condition). We also suppose that the rate of quark production is somewhat higher than that of antiquarks (the second condition). If the reverse processes of quark or antiquark destruction are suppressed (the third condition), the specified interaction will naturally produce an excess of matter over antimatter in the universe.

This general scheme of baryogenesis looks very plausible. Researchers can test it and try to uncover that special force of nature which would satisfy all three conditions and make the dominance of matter possible. In the remaining part of the book, we will follow the progress of scientists in this search. But before proceeding in the indicated direction, we need to clarify one important point. The second condition requires there to be some difference between matter and antimatter. However, it does not stipulate that the unknown interaction completely forbids the production of antiparticles. In fact, none of the realistic theoretical models assume such an overwhelming advantage of matter over antimatter. On the contrary, for each process creating an excess of quarks, all models expect a corresponding process generating a surplus of antiquarks. The second condition means that the number of antiquarks produced should be somewhat lower. However, the difference between particles and antiparticles could be tiny.

After saying this, we need to admit that the total number of antiquarks created after the Big Bang must have been huge, even if nature preferred matter to antimatter. If so, why did all that antimatter disappear?

Physics can already answer this question. We don't observe antimatter because all antiparticles have annihilated together with the same number of particles. But this is just one part of the answer. The energy of particles that filled the universe soon after the Big Bang was very high. Therefore, annihilation copiously produced secondary quarks and leptons together with corresponding antiparticles in processes similar to those shown in Fig. 5.6. Subsequently, the secondary fermions and antifermions also annihilated. The mutual destruction continued until the energy of created objects was reduced below a certain threshold, such that no new particles and antiparticles could be produced anymore. Afterwards, the only possible outcome of annihilation was photons, neutrinos, and antineutrinos.

If matter did not have any advantage over antimatter, the end of the story would be rather sad. It is true that a small number of charged fermions and antifermions would survive even in this case. When the density of particles and antiparticles would become small enough, they would not collide anymore. It is also true that the remaining quarks and antiquarks would be able to produce baryons and antibaryons, respectively. However, the amount of the remaining matter would be billions of times less than we now observe. In any case, particles and antiparticles would not gather into stars or antistars because any substantial concentration of matter and antimatter would cause new waves of mutual annihilation. Hence, that would be a dark and empty universe. But nobody would care because life would not be possible in that world.

Conversely, since Sakharov's second condition is satisfied, the destiny of the universe will be different. The number of quarks created slightly exceeded the number of antiquarks. The chain of annihilations eliminated equal quantities of quarks and antiquarks. But a small residual number of quarks persisted, they did not disappear because no antiquarks existed anymore to destroy them. So, the remaining quarks had no choice but to combine into baryons. Baryons united with electrons and produced atoms. Atoms were pulled together by gravitational force. The density of matter in some parts of the space became so high, that the first stars were ignited. Gradually, the universe took the familiar shape of billions of shining galaxies embedded into the vast space.

Thus, the proposed model of baryogenesis can definitely explain why we currently don't detect antimatter. This model assumes that all matter, which we now observe in the universe, is just surviving residue after primordial annihilations have exhausted themselves. A much greater number of particles and antiparticles were created and, almost simultaneously, destroyed in the first few instants after the Big Bang.

The proposed model is not just a theoretical deliberation. It can be experimentally verified. It turns out that the total number of photons, which currently fill the universe, is approximately the same as the number of photons created during the primordial annihilations. So, this quantity should correspond to the number of fermions and antifermions which annihilated many billions of years ago. Of course, the proportions of photons and initial quarks cannot be exactly the same due to many reasons. But a factor of three or four is not essential for the following discussion. What is important is that if the proposed model of baryogenesis is correct, the number of baryons in a unit volume of space (that is, the residual excess of matter over antimatter) will be much smaller than the number of photons (that is, the initial number of produced particles and antiparticles). Hence, the ratio of these two numbers is an important quantity. It quantitatively reflects the excess of matter over antimatter during baryogenesis.

The ratio of the number of baryons to the number of photons in a unit volume can be extracted from the measurement of the *cosmic microwave background*, or *CMB*. This term denotes the photons left over from the early stages of the universe's evolution. Initially, when the universe was hot, protons and electrons had high energy and existed as isolated objects. Consequently, photons could not propagate to large distances because of permanent interactions with these charged particles. Figuratively speaking, the universe was murky and dark. However, when the temperature of the universe decreased, baryons and electrons slowed down and combined into atoms. At that time, the frequency of photon interactions was considerably reduced because atoms are neutral objects. Hence, the universe became transparent, and photons could freely move in all directions. These photons continue to exist up to the present days and reveal itself as microwave radiation, or CMB, which fills all the space of the universe. The CMB can be detected either on Earth or using radio telescopes installed on board space satellites.

The space observatory COBE launched by NASA in 1989 performed one of the first precise measurements of the CMB. COBE was followed by the experiment WMAP, which reported the first measurement of the ratio of baryons to photons. The most precise measurement of this ratio was obtained by the space observatory Planck. The observatory and the corresponding satellite were named after the German theorist Max Planck, who made a significant contribution in the development of the theory of photons.

The Plank spacecraft, launched in 2009, was placed into the so-called L_2 Lagrange point of the Sun-Earth system. This point is located beyond the orbit of Earth on the line joining the Earth and the Sun. There are five Lagrange points in the Sun-Earth system. Objects placed in the Lagrange points don't change their position with respect to the Earth, even though our planet moves around the Sun. The particular advantage of the L_2 point is that the Earth partially screens a satellite placed at this point from solar radiation. This screening simplifies the calibration and adjustment of the instruments of the PLANK observatory. Also, the background electromagnetic radiation produced by our civilization is substantially reduced at the L_2 point, since the distance to the Earth's surface from this point is about 1.5 million kilometers.

The precise measurement of the CMB allows scientists to get several important quantities, such as the age of the universe and the fraction of dark matter in it. The ratio of baryons to photons is also one of the parameters extracted. According to the latest results, for every two billion photons present in the universe, there is just one baryon. Although here on Earth the amount of matter is enormous, at the scale of the universe, on average, one cubic meter of the space contains only about six baryons while the number of photons in the same volume is considerably higher.

The obtained result is equivalent to stating that for every billion of regular quark-antiquark pairs produced after the Big Bang, there was just one additional quark created without an accompanying antiquark. Subsequently, myriad quark-antiquark pairs have annihilated, but the orphan quarks have survived. This seemingly negligible remainder has turned out to be sufficient to produce all the richness of the universe, which we can now admire when gazing out at the night sky.

Thus, regardless of the present overwhelming predominance of matter, its advantage in the fight against antimatter was extremely small. It was not ten to nine or even 100 to 99. The surplus was just one part in a billion. This result is remarkable because it makes the proposed scheme of baryogenesis highly credible. It seems reasonable to expect such a small excess of particles over antiparticles, and theoretical models can try to accommodate it. Hence, the measured ratio of baryons to photons supports and motivates further theoretical and experimental research in this direction. So, keeping this result in mind, we can now resume the story of antimatter.

Chapter 14

Searches for Baryon Number Violation

14.1 Proton Decay

The general scheme of baryogenesis presented in the previous chapter is based on Sakharov's conditions. Their formulation was an important step forward in understanding the disappearance of antimatter. However, just stating the requirements is not sufficient. Researchers need to find actual mechanisms putting the conditions into effect. In this respect, the first condition, which specifies the violation of baryon number, seems the most puzzling, because it has never been observed in nature.

The baryon number violation would manifest itself, for example, as the creation or decay of a baryon without an accompanying antibaryon. Decays of baryons are well-known. For example, an isolated neutron decays to a proton, an electron, and an antineutrino. But the baryon number is not violated in such decay because a proton appears instead of a neutron. On the contrary, the decay of an isolated proton would be a clear sign of the baryon number violation. As was discussed earlier (see Section 9.2), the mass of the decaying particle must always be higher than the mass of the decay products. The proton is the lightest existing baryon. Consequently, no other baryon can emerge instead of the decaying proton, and the baryon number would decrease by one unit after *proton decay*. That is why the observation of this decay would be an unambiguous signature of baryon number violation. It should be stressed that this violation can reveal itself not just in proton decay. There are several other possibilities to detect it. Moreover, the baryon number can be violated even if a proton is a truly stable particle and never decays. However, proton decay is the most convenient process to reveal this violation from the experimental point of view. So, it is not surprising that attempts to find proton decay are currently among the most active directions of research, which aims to prove Sakharov's first condition.

If the instability of protons is indeed uncovered, it will considerably impact the entire understanding of the formation of the universe and its ultimate destiny. Of course, the prospect of proton decay does not mean that all protons of the universe will instantly disappear. However, their vanishing will be inevitable. All processes in the microworld are probabilistic. Nobody can tell how long a certain particle will exist. Only the *mean lifetime* of particles makes sense. For each type of particle, this quantity is defined as the period, after which only 37% of an initial sample of the particles remain. All other particles in the sample should decay by that time. After one more period, 63% of the surviving particles will also decay, and just 13% of the initial sample will remain. This process will continue until all particles disintegrate.

Thus, if particles of a certain type can decay, their complete disappearance is just a question of time. Therefore, admitting proton decay, we must also accept that "diamonds are not forever", as expressed by Sheldon Glashow, one of the founders of the Standard Model. In other words, atoms made of protons and neutrons are not everlasting and will cease to exist sometime in the future. In its turn, the destruction of atoms will signify the end of the universe in its current form. The death of the universe will not be instantaneous, as it happens with humans and animals. It will be a slow extinction. The material of stars will gradually diminish like smoke on water, and the stars will be extinguished one after another until all space becomes empty of atomic matter. Only radiation will finally remain as a silent witness of the former greatness of the universe.

There is some comfort despite this gloomy picture of the apocalypse. The mean lifetime of a proton cannot be too small. At least, it cannot be less than the time passed after the Big Bang. Otherwise, a substantial part of the primary protons would have already decayed, and the consequences of this destruction of matter would be visible. Regardless of the specific modes of proton decay, this process would release energy stored in the proton's mass. A substantial part of this energy would transform into radiation, which can be detected here on Earth. But such radiation was never observed. Therefore, protons are stable enough. The age of the universe is estimated to be about 14 billion years. Consequently, the mean lifetime of a proton cannot be smaller than this value. Thus, we should not worry yet about the end of the world, at least for the next few billion years. Hopefully, this is a sufficient time for the human civilization to prosper.

If proton decay is possible, it must proceed through some interaction. The nature of this interaction is hard to discern. First, there is not any confirmation of its existence, and it is always frustrating to study something that nobody has ever observed. But more importantly, the properties of this interaction must be substantially different from anything else. The modes of proton decay are quite limited. Some mesons, like a pion or a kaon, have mass smaller than that of a proton. However, all mesons are unstable and decay sooner rather than later. The only stable particles that could remain after proton decay are photons, electrons, and neutrinos, together with their antiparticles. The charge of a proton is positive, and the sum of charges of all decay products must also be positive. Consequently, the decay of a proton must always produce a positron, which is the only suitable particle with a positive charge. A proton is made of quarks, while a positron is an antilepton. Thus, the interaction governing proton decay must perform a very unusual transformation. It should convert a quark into an antilepton.

Possible arrangements of such an abnormal transformation can be clarified by revising what we already know about the familiar forces of nature and their action on particles. Electromagnetic interactions always leave the type of interacting particles unchanged. Strong interactions can change the color state of a quark, but do not alter the quark flavor. Weak interactions are the most sophisticated. They can modify the flavor of fermions. But if we treat two different flavors as two *flavor states* of the same particle then the action of weak interactions becomes quite similar to that of strong interactions. For example, the transformation of a down quark into an up quark can be considered as a transition of a quark from one flavor state (down) to another (up). Similarly, when an electron turns into an electron neutrino, we can say that the flavor state of a particle (lepton) changes from electron to electron neutrino. In this interpretation, weak interactions do not change the type of a particle (lepton or quark), but only modify its flavor state. Of course, an electron is very different from a neutrino. However, there are many examples when two entirely different entities are, in fact, two states of the same thing. For example, a caterpillar, the butterfly's larva, and a butterfly do not look alike at all. Still, they are just two states of the same organism.

This discussion leads to an interesting idea. Hypothetically, the interaction responsible for proton decay can be arranged in such a way that all quarks and leptons together with their antiparticles are perceived by this interaction as different states of a *single super-particle*. Here again, the analogy to an inhabitant of a house with several rooms, can help. A person eats in one room, sleeps in another, and watches TV in the third. But regardless of these apparently different activities, the person always remains the same. Following this analogy, we can treat an electron or an up quark as different "activities" of the same super-particle. In this case, the transformation of a quark into an antilepton can be presented as a change of the state of the super-particle. Consequently, a theory describing these unusual transformations can be like the well-developed theories of strong and electroweak interactions.

The proposed idea of a super-particle seems very productive because the theoretical models that implement it are numerous. One of them is considered in more detail in the next section.

14.2   Grand Unification Theory

One of the most popular models that describe the unification of fermions into a single super-particle is called SO(10). Such an odd

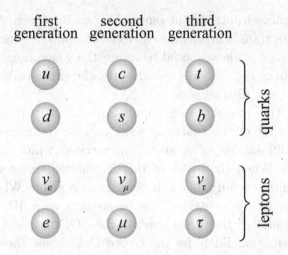

Fig. 14.1   Three generations of fundamental fermions.

name has a solid mathematic background, which is nevertheless beyond the scope of this book and is not discussed here.

All fundamental fermions can be divided into three groups called *generations*, which are presented as columns in Fig. 14.1. The SO(10) model treats all fermions and antifermions of one generation as different states of a single super-particle. More precisely, the super-particle of the first generation includes up and down quarks, an electron neutrino, an electron, and the corresponding antiparticles. Another super-particle combines heavier quarks and leptons of the second generation: charm and strange quarks, a muon neutrino and a muon plus their antiparticles. Finally, the heaviest fermions (top and bottom quarks, a tauon neutrino, a tauon, and their antiparticles) constitute the third super-particle.

The division into generations is not arbitrary. All interactions between the particles of the first generation are almost exactly reproduced in the second and third generations. Also, the rate of quark transformation to another quark of the same generation is much higher than to the quarks of other generations. For example, the transition of charm to strange quark is much more probable than to down quark. An up quark transforms much more often into a down quark than into a strange quark. Moreover, the transitions between leptons from different generations are forbidden. An electron neutrino can transform only into an electron and can never become a muon. Because of these special relations, the generations are clearly separated from each other. Consequently, the super-particles can be treated as three distinct objects.

The SO(10) model not only unites quarks and leptons but also treats three interactions (strong, weak and electromagnetic) as different manifestations of a single interaction, which governs the microworld. When the boson of this interaction transforms a blue up quark into a red up quark, it behaves as a gluon. When it turns an electron into a neutrino, this boson acts as a W boson and so on. Because of this arrangement, the SO(10) model is considered a possible candidate for the *Grand Unification Theory (GUT)*. Many generations of scientists have dreamt of such a theory, which would unite all known interactions and explain all phenomena of the microworld using a universal approach. In this respect, the SO(10) model represents a very promising attempt to realize the dream. The model successfully describes all observed phenomena of particle interactions using a uniform theoretical basis. But it also predicts something new.

The SO(10) predictions include the transformations of quarks into antiquarks or antileptons. The model treats all such unusual transitions as changes of the state of the super-particles. Hence, the arrangement of these transformations is comparable to the well-known transitions of quarks and leptons. In particular, the SO(10) model contains special bosons that mediate the conversion of quarks into antiquarks or antileptons, and cause proton decay.

The diagram corresponding to proton decay is shown in Fig. 14.2. In the specified process, the up quark is transformed into an up antiquark while the down quark turns into a positron. The boson of the SO(10) model, which conducts the unusual transitions, is denoted by letter X. Two arrows on each fermion line point in opposite directions. It is a clear violation of previously specified rules for Feynman diagrams. But this feature just reflects the unusual properties of the transitions conducted by the X boson. The diagram in Fig. 14.2

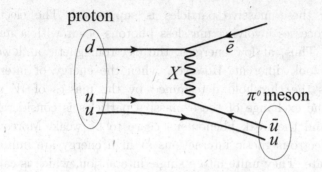

Fig. 14.2   Proton decay in the SO(10) model.

implies that the primary decay mode of the proton should be a positron and a neutral pion.

Notwithstanding all attractive features of the SO(10) model, none of its predictions beyond the results of the Standard Model are experimentally confirmed. However, there is a plausible explanation to this limitation. It turns out that the mass of the X bosons exceeds by many billion times the mass of all known particles. If we recall the analogy of interaction and a football game, the exchange of X bosons can be compared to the game with an iron ball. Normal players would not kick such a ball without injuring themselves. However, giants the size of a mountain would easily do this. Similarly, the interacting particles must have a very high energy to exchange X bosons at a reasonable rate. Conversely, if the energy of interacting particles is lower than the threshold of the X mass, the effects produced by X bosons will be negligible. The level of energy required for X exchange is much higher than the energy achieved in contemporary collider experiments. That is why exotic interactions involving X bosons have never been observed.

Massive bosons are not something unusual in the microworld. For example, both the W and Z bosons, which determine weak interaction, have high masses. These masses are much greater than the masses of most leptons and quarks. When the energy of interacting particles is lower than the.mass of W and Z bosons, the exchange of these massive particles is suppressed. The electro-magnetic processes involving massless photons occur with a much higher rate. Thus, at low energies, the electromagnetic and weak interactions look different. However, when the energy of interaction exceeds the threshold determined by the masses of W and Z bosons, the exchange of these massive particles is considerably simplified, and the weak phenomena cease to be weak. Moreover, weak and electromagnetic interactions at high energy are impossible to separate. They unite into a single interaction, which is called *electroweak*.

A similar effect is expected in the SO(10) model. The only difference is that the bosons of the GUT model have a mass billions of times higher than W and Z bosons. At low energies, all manifestations of the SO(10) model beyond well-known electromagnetic, weak and strong effects are invisible, and the rate of unusual transitions, such as quarks into antileptons, is strongly suppressed. However, when energy approaches the threshold of the X mass, the three forces of nature become comparable and start to act as different forms of a single interaction. Simultaneously, the rate of unusual transitions increases manifold. The threshold of energy, above which the electromagnetic, weak and strong forces are united, is called the *threshold of grand unification*. This threshold roughly corresponds to the mass of the X boson.

It can be suspected that theorists intentionally choose a high value of the threshold of grand unification to justify the absence of visible manifestations of the SO(10) model. However, this threshold is not a free parameter. Its value is entirely determined by the requirement that all interactions unify. Therefore, the absence of new effects is, in fact, one of the predictions of the GUT model.

The energy of particles involved in proton decay is determined by the rest energy of a proton. This energy is much lower than the threshold of grand unification. Consequently, the rate of proton decay must be strongly suppressed, and a proton should live a very long life. When expressed in years, the mean lifetime of a proton is represented by a number of 35 digits. For comparison, the time passed after the Big Bang is a number of 10 digits (about 14 billion years). Thus, the number of protons which have decayed during this time is negligible, and protons can be treated as stable particles. Nonetheless, the fact that protons can decay implies baryon number violation, which is essential for explaining the disappearance of antimatter. That is why the prediction of proton decay is one of the key achievements of the SO(10) model.

It may seem that the extremely long lifetime of the proton contradicts the general scheme of baryogenesis outlined at the beginning of the previous chapter. All processes leading to the predominance of matter over antimatter must have ended long before the first second of the universe's life. But the baryon number violation is so suppressed that a proton cannot decay even after billions of billions of billions of years. Apparently, such slow processes would not produce any visible impact on the struggle between matter and antimatter and would not contribute in the establishment of the predominance of matter in the universe.

The solution to this contradiction is the following. Immediately after its birth, the *temperature of the universe* was very high. This temperature is determined by the energy of radiation, which fills the space of the universe. Radiation interacts with other particles and transfers a part of its energy to them. Therefore, the temperature of the universe is a quantitative representation of the energy of all particles in the universe. Currently, the temperature of the universe is just 2.73° above absolute zero on the Celsius scale.[1] But it wasn't always like this. The temperature of the universe was billions of times higher immediately after the Big Bang. It was so high that the energy of all particles was comparable to the threshold of grand unification and even exceeded it. Consequently, the interactions involving X bosons proceeded with a very high rate. Figuratively speaking, for the particles possessing the energy of giants, the exchange of massive X bosons was like a child's game. This exchange resulted in the transitions of quarks into antileptons, and such transitions implied baryon number violation. The rate of such processes was sufficiently high, so that nature had enough time to produce a small but extremely

---

[1]For comparison, the temperature of water freezing is 273° above the absolute zero on the Celsius scale.

crucial excess of matter over antimatter. Thus, the theory of grand unification, and, in particular, the SO(10) model, provides a reasonable explanation for why interactions involving X bosons were substantial at the time of creation of matter in the universe, but are not observed today.

In addition to the SO(10) theory, scientists have developed many other models with similarly odd names. All of them are considered candidates for GUT. All of them have comparable structures and predict proton decay. However, the numerical values of the mean lifetime and the decay modes of the proton predicted by various models are different. For example, some models claim that the dominant decay mode of the proton is a charged kaon and a neutrino.[2] Thus, the discovery and detailed study of proton decay would not only validate the main principles of GUT, but it would also considerably reduce the list of candidates for the title of the Theory of Grand Unification. What is also important is that the discovery of proton decay would confirm baryon number violation. This proof would be a decisive success in explaining the disappearance of antimatter.

Experimental results related to the search for baryon number violation are discussed in the next section.

14.3   Experimental Search for Baryon Number Violation

The detection of proton decay appears to be Mission Impossible. Indeed, it seems meaningless to wait for the decay of a particle, if even the entire lifetime of the universe is not enough for this to happen. However, scientists are quite positive that this can be done. Even a single gram of matter contains a huge number of protons and neutrons. Hence, proton decay ought to be discovered by observing a significant amount of any material for a sufficiently long time. The effect will be established even if just one proton decays during the

---

[2]The charged kaon is unstable, and its ultimate decay products include a positron, neutrinos, and antineutrinos. Therefore, a positron is always produced in proton decay.

time of observation. It is like playing the lottery. Each person has a negligible chance to win the lottery jackpot. Still, if the number of players is sufficiently large, somebody will succeed.

The best result so far in the search for proton decay has been obtained by the Super-Kamiokande experiment in Japan. In this study, scientists monitor a tank of ultra-pure water. The reservoir consists of a cylinder 40 meters tall and 39 meters in diameter. Thus, the total mass of water used in the experiment is about 50,000 tons. The method of proton decay detection is quite simple. It is based on a well-known phenomenon, *Cherenkov radiation*, discovered by the Soviet scientist Pavel Cherenkov. By itself, it is a fascinating effect caused by particles moving faster than light. Of course, nothing can move faster than light in a vacuum. However, the speed of light in any medium is lower than the speed of light in a vacuum. On the other hand, even in a medium, the speed of a massive particle is limited only by the speed of light in a vacuum. Therefore, a sufficiently energetic particle moving in a medium can outstrip light. In such a case, the charged particle emits electromagnetic radiation.[3] Hence, registering this radiation in some volume of water would be an unambiguous indication of an energetic particle crossing the medium. Cherenkov radiation is mostly emitted as ultraviolet light, but part of it is visible to the eye as a characteristic blue glow. So, the detection of this radiation is relatively simple.

Suppose that a proton decays somewhere in the volume of the water tank used in the Super-Kamiokande experiment. As discussed earlier, the final state of the process must include a positron. This positron will travel a considerable distance before interacting with particles of the medium. The mass of a proton is large, and a substantial fraction of its rest energy goes to the positron. A positron, in its turn, is a very light particle, so the energy transferred is sufficient to accelerate the emitted position to the speed that exceeds the speed of light in water. Therefore, the moving positron will

---

[3]Neutral particles cannot do this because they do not interact electromagnetically.

emit Cherenkov radiation. Consequently, proton decay should manifest itself as a sudden flash of ultraviolet light accompanied by a blue glow.

The water used in the Super-Kamiokande experiment is ultrapure. Hence, radiation emitted somewhere in the bulk of water would fade very slowly and propagate up to the edges of the tank. There, it will be registered by photodetectors installed on the reservoir walls. In this setup, the energy of the positron produced and its direction of motion can be determined. Even the point of origin of the positron can be established.

The Super-Kamiokande experiment started in 1996. Since then, the scientists have continuously hunted for proton decay, day after day and night after night. This hunt was not seamless as several incidents happened during the extended period of the experiment running. The most serious one occurred in 2001. One of the photodetectors was broken during a routine maintenance work. A photodetector used in the experiment consists of a vacuum glass tube about 50 centimeters in diameter. When such a tube was broken, it imploded, and water filled the space inside the tube in a fraction of a second. If the break had happened in air, it would not have caused any big problem. But the break occurred at a depth of 40 meters under the water surface. Because of high pressure, the flow of water that filled the broken photodetector was violent. It produced a shock wave, which quickly propagated to all parts of the detector. The power of this shockwave was so strong that other photodetectors began to implode one after another. Because of the chain reaction, almost half of all the photodetectors were destroyed in less than a minute.

It was a considerable loss for the experiment since the cost of photodetectors was high. However, the search for proton decay was pursued despite a substantial reduction in efficiency. The remaining photodetectors continued to be used to watch the darkness of the water. The scientists kept on with monitoring their readings. They never abandoned hope of seeing the flash of blue light so long dreamed of. All broken photodetectors were replaced in 2006. Afterwards, the experiment resumed at its full capacity. However, no positive results have been achieved so far. The scientists have only succeeded in obtaining the *lower limit* on the mean proton lifetime. That is, they can confidently claim that the mean proton lifetime cannot be less than a certain value. The lower limit obtained is very large. It is expressed as a number of 34 digits. However, it is still far from theoretical expectation, which is represented by a number of 35 digits in the SO(10) model.

The gap between the experimental limit and the theoretical expectation may seem quite small. But in fact, it is huge. The Super-Kamiokande experiment has already been working for 20 years. To add one digit to the lower limit, it needs to work about 200 years more. Nobody expects that the investigation will continue for 200 years, and nobody can live for such a long time. The only alternative to a virtually infinite waiting time is an increase in the mass of material used to search for proton decay. There are several proposals for new experiments which follow this direction. One of them is called Hyper-Kamiokande. The scientists of this experimental proposal hope to increase the mass of water by 20 times. With such improvement, proton decay should be observed during a human's lifetime provided the theoretical estimate is correct.

Unfortunately, the theoretical prediction is not stable and can easily be modified in one direction or another. At least, such modifications have happened previously. All models of GUT agree that the threshold of grand unification must be extremely high, but its actual value can vary by as much as an order of magnitude. Consequently, the expected lifetime of the proton can also change. Initially, theoretical models predicted a reasonably small value of the mean proton lifetime so that the experimental detection of proton decay was achievable. However, after experimenters attempted to do this and failed, the theoretical model was adjusted, and the expected mean lifetime was increased.

In any case, playing catch-up does not diminish the interest of experimenters in proton decay. They continue their attempts to observe this fascinating phenomenon, which is of immense importance for the future development of physics.

In addition to proton decay, there is one more possible manifestation of baryon number violation. It is the spontaneous transformation of a neutron into an antineutron, which is also called a *neutron oscillation*. A proton cannot turn into an antiproton because this transition would violate charge conservation. However, a neutron is a neutral particle. None of the fundamental laws forbid its conversion into an antineutron. The transformations of a particle into its own antiparticle are well-known (see Chapter 11). Such transformations happen spontaneously, without the influence of external factors. In all known transformations of this kind, a quark-antiquark pair converts into another quark-antiquark pair, as in the diagram shown in Fig. 11.1. Therefore, the baryon number is not violated in these processes. On the contrary, the neutron consists of three quarks. If it converts into an antineutron, the baryon number will change by two units. Consequently, the observation of this transition will be an undeniable confirmation of Sakharov's first condition.

Similar to proton decay, the search for neutron oscillation is based on the principle of winning the lottery. Any material contains a huge number of neutrons. For example, a molecule of water contains an atom of oxygen. This atom contains eight neutrons, and the number of molecules in one gram of water is of 23 digits. Such a big number of neutrons increases the chances of detecting the neutron oscillation to an acceptable level.

Every particle and its antiparticle has exactly the same mass and, consequently, the same rest energy. Therefore, the transformation of a neutron into an antineutron would not be accompanied by the release of energy or by the generation of new particles. However, the created antineutron would immediately annihilate with other baryons of the atomic nucleus. This annihilation would produce a large number of pions and other mesons. All these particles are very energetic. The charged pions would live long enough to travel a substantial distance before the disintegration. Consequently, they also should emit Cherenkov radiation when they move in water. Hence, the search for neutron oscillation is technically quite similar to the quest for proton decay. In both cases, the expected effect would reveal itself as a sudden flash of Cherenkov radiation in a bulk of water. That is why the physicists of the Super-Kamiokande experiment explore both possibilities of baryon number violation. The upper limit on the likelihood of neutron oscillation obtained by this facility is currently the world's best. The only problem is that it is still a limit and not an observation of the phenomenon.

Regardless of the negative result, the search for neutron oscillation will continue. The possibility of baryon number violation is one of the most important questions which should be addressed by particle physics. In this respect, there is an interesting proposal for a new experiment called NNbarX. It has been planned to be conducted in the scientific center Fermilab in the USA. Contrary to the Super-Kamiokande, the NNbarX experiment will concern itself with the search for the oscillation of neutrons produced by an accelerator. The modern accelerator technologies allow for the production of a very intense beam of low-energetic neutrons. Such a beam will be directed into a special vacuum pipe. The speed of neutrons will be relatively small so that they will travel from one edge of the pipe to another in about half a second. If one of the neutrons spontaneously turns into an antineutron, the created antiparticle will continue to move in the same direction together with other particles. On exiting the pipe, the antineutron will annihilate in the material of a detector, which will be located immediately where the pipe ends. This annihilation will be registered, and, consequently, the phenomenon of neutron oscillation as fact will be established. Preliminary estimates show that the NNbarX experiment can considerably improve the current limit of Super-Kamiokande. But nobody can guarantee that neutron oscillation will be detected.

14.4 Baryon Number Violation in the Standard Model

Baryon number violation is an important prediction of several models of Grand Unification. All these theories are prone to an apparent difficulty: the limited prospects of their tests. A theory that cannot be tested becomes like a religion because people have no choice but to believe in it. To avoid this destiny, the GUT models need to get at least one confirmation of the unusual particle properties which they predict. Currently, this is not attainable because the mass of the bosons that determine the unusual interactions is much higher than the energy achievable at all current and future colliders. Hence, the prospects of discovering these bosons or at least, detecting their effects are bleak.

In this respect, there is another interesting possibility of baryon number violation which does not require new theories. This opportunity is present in the Standard Model itself. About fifteen years after the creation of the Standard Model, theorists realized that this theory implies the existence of unusual processes, which were named *sphalerons*. The word "sphaleron" was derived from the Greek word meaning "slippery", but such translation does not clarify the meaning of the phenomenon at all.

Sphalerons are completely different from all other effects that happen in the microworld of particles. The long time needed to realize the possibility of these processes underlines their singularity. They are not related to an exchange of bosons, and this feature alone makes sphalerons unusual. All interactions considered so far involve four fermions at most. On the contrary, sphaleron processes require a simultaneous participation of at least twelve quarks and leptons. Moreover, all fermions must be from three different generations. For example, a sphaleron process can transform two quarks into three

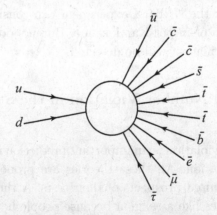

Fig. 14.3   Example of a sphaleron process.

antileptons and seven antiquarks. The picture corresponding to this transformation is shown in Fig. 14.3. In general, sphaleron processes cannot be presented by Feynman diagrams since the usual bosons do not conduct them. Consequently, pictures like those in Fig. 14.3 do not follow any specific rules and are not too informative. They just give a graphical presentation of transformations, which happen during sphaleron processes.

A general feature of all sphaleron processes is that they simultaneously violate both baryon and lepton number conservation. However, the difference between these two numbers always remains the same. For example, let us consider the process shown in Fig. 14.3. Each quark has 1/3 of the baryon unit because each baryon contains three quarks. Similarly, each antiquark has −1/3 of the baryon unit. Consequently, the difference between the initial and final baryon number is equal to −3 (two quarks in the beginning and seven antiquarks in the end). Simultaneously, the lepton number is decreased by three units because three new antileptons are created. So, the difference between the baryon and lepton number remains the same at the beginning and the end of the sphaleron process. Going forward, we need to stress that this property of sphaleron processes will be essential for the scheme of baryogenesis discussed later in Chapter 19.

Sphaleron processes happen only when the energies of participating particles are very high. The energy threshold considerably exceeds the possibilities of current collider experiments. So, observations of sphaleron processes are still impossible. But this threshold was not an obstacle for baryon number violation in the first few instants after the Big Bang. At that time, the temperature of the universe was very high, and interacting particles had the energy required.

Limitations in observing sphaleron processes may seem similar to those of the GUT models. However, there is one significant difference between them. According to theoretical calculations, the energy required for sphaleron processes is much lower than the threshold of energy for grand unification. Therefore, baryon number violation by sphaleron processes can happen at a much lower temperature of the universe in comparison to the requirements of the GUT models. Consequently, when the temperature of the universe decreased, the sphaleron processes were still active for a long time after the end of effects predicted by GUT models. We will see later that this feature is also essential for baryogenesis.

Sphaleron processes are predicted by the Standard Model. So, they must exist as far as this theory is valid. The Standard Model, contrary to the GUT, has been thoroughly verified by numerous experiments. In all such tests, the predictions of the Standard Model have always been confirmed. The latest of such confirmations, the discovery of the Higgs boson, was obtained just a few years ago. Hence, the spotless "credit history" of the Standard Model provides the necessary confidence in the possibility of sphaleron processes.

Since the energy required for sphaleron processes is lower than the threshold of grand unification, the detection of sphalerons is a more realistic expectation than that of the GUT bosons. For example, the sphaleron processes could be generated by highly energetic neutrinos coming to the Earth from the depths of space. In some cases, the energy of these neutrinos is considerably larger than the energy of particles in collider experiments. Such neutrinos can interact with detector material, and sphaleron processes can be searched for among these interactions. The possibility of an experiment that would detect such effects is now actively discussed in scientific papers. Also, sphaleron processes can be explored at the LHC in CERN, or at more energetic colliders in the future. These are currently planned.

Such is the status of the search for baryon number violation. This search goes on in several directions. Firstly, the phenomenon is predicted by several promising models, which contest the title of Grand Unification Theory. According to these models, baryon number violation can manifest itself as the decay of a proton or the oscillation of a neutron. Secondly, sphaleron processes, expected in the Standard Model, should also violate baryon number conservation. Up to now, the search for baryon number violation has not brought any positive result. But the absence of this violation here and now does not prevent it from occurring during the initial stages of the universe's evolution. According to all theoretical models, the rates of the processes leading to baryon number violation considerably increase with the increase in energy of interacting particles. Therefore, baryon number violation is expected to have frequently happened soon after the Big Bang, when the universe was very hot. Because of this violation, antimatter has ultimately disappeared from our world.

Chapter 15

CP Violation

15.1 CKM Matrix

Contrary to Sakharov's first condition, the existence of CP violation has been confirmed by numerous experimental measurements. The effect was first detected in the decays of neutral kaons more than 50 years ago. Since then, CP violation has been actively studied in processes involving other hadrons, and many other manifestations of this phenomenon have been discovered. These explorations have uncovered a particular mechanism of CP violation, which is responsible for all known effects that violate CP symmetry. This mechanism is important for our story because an understanding of CP violation should help in explaining the disappearance of antimatter from the universe. That is why the achievements in the study of CP violation are presented below in more detail.

Weak interaction governs the decays of neutral kaons. Hence, this type of interaction possesses some special features that cause CP violation. To identify these features, let us consider more attentively the diagrams in Fig. 9.6. They correspond to the decays of a neutral kaon (left) and antikaon (right) into two oppositely charged pions. Such decays were studied in Cronin and Fitch's experiment. At first glance, the two diagrams look very similar. Still, there is one principal difference between them. In the decay of a kaon (left plot), the quark transitions are[1] $u \to s$ and $d \to u$. In the decay of an antikaon (right plot), the transitions are performed in the reverse order: $s \to u$ and $u \to d$. This observation reflects a general rule that relates decays of any hadron and its corresponding antihadron. All differences between them boil down to the order of quark transitions. If the decay of a hadron is caused by the transformation of a certain quark x into quark y, then the decay of an antihadron is determined by the reverse transformation[2] of the quark y into quark x. This rule leads to an exciting conclusion. If the transition $x \to y$ differs from $y \to x$, properties of the corresponding hadron and antihadron will be different, and CP symmetry will be violated in their decays.

Thus, once again, a fundamental conclusion can be derived just by looking at Feynman diagrams. We can expect that **differences between direct and reverse quark transitions generate CP violation**. This statement agrees well with the previous result that CP violation is equivalent to T violation (see Section 10.3). Indeed, the specified difference corresponds to the violation of the symmetry of time reversal. At the same time, this difference results in the distinction between a hadron and antihadron. That is, it produces CP violation.

The difference between direct and reverse transitions is possible only because fermion flavors are changed in weak interactions. In all other interactions, the flavors are always preserved, and, consequently, direct and reverse transitions coincide. For example, the interaction of an electron with a photon corresponds to the transition $e \to e$, which remains the same after time reversal. Hence, CP symmetry must be conserved both in this and in all other electromagnetic interactions. Indeed, all experimental results confirm this conclusion. CP violation has not been detected in strong interactions either. This observation is also explained, among other reasons, by

---

[1] As usual, to determine the type of quark transition, we need to follow the arrows on the fermion lines.

[2] Of two transformations, $x \to y$ and $y \to x$, one will be called direct and another reverse. Of course, specifying what is direct and what is reverse is completely arbitrary.

the fact that the flavors of quarks remain unchanged in the interactions with gluons.

Weak interaction represents the only known source of CP violation. As we discussed earlier, without CP violation, the universe would contain matter and antimatter in equal proportions. Particles and antiparticles would continuously collide and annihilate. Consequently, without weak interaction, the structure of the universe observed would never materialize.

The conclusion on the importance of weak interactions for the evolution of the universe to its current form is quite surprising. The atomic structure of matter is determined mainly by the strong and electroweak forces while weak interaction may seem auxiliary and inessential. But on the contrary, we now see that it plays a central role. Nature can be compared to a finely tuned mechanism, where all parts work in strict conformity with each other. No details are redundant, and their joint action determines the final result.

The above discussion underlines the importance of fermion flavor transformations for understanding CP violation. That is why physicists pay them a great deal of attention. The most striking feature of these transformations is that **the transitions between different quarks are quantitatively inequivalent**.

Suppose that we study collisions of an antineutrino with an up quark. If the energy of the antineutrino is high enough, the up quark can transform into a down, strange, or bottom quark. The corresponding diagrams are shown in Fig. 15.1. These diagrams are quite similar. Essentially, the only difference between them consists of replacing the symbol of one quark with another. This observation is pertinent because Feynman diagrams not only give a graphical presentation of processes but also allow for a calculation of all their

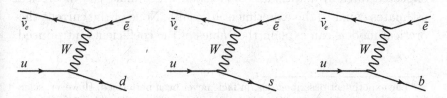

Fig. 15.1   Interaction of antineutrino with up quark.

properties. As can be expected, calculations based on similar diagrams should give close results. In the case of Fig. 15.1, this means that the production of the three flavors can be different only because of the variation of their masses. The mass of a bottom quark is about a thousand times larger than that of a down quark. Still, if the energy of the antineutrino were very high, we would expect that the influence of mass would diminish, and the production rate of all quark flavors would be almost identical. However, if we perform the actual measurements, the rates will be significantly dissimilar. A down quark will appear in most cases. The probability of observing a strange quark will be about 5%. A bottom quark will be produced only in one out of 81000 cases.[3]

The rates of all other quark transitions are also significantly different. The scale of this disparity is graphically presented in Fig. 15.2. The area of each square in the figure is proportional to the rate of transition between the corresponding quarks. The numerical inequality of the quark transitions highlights the division of all quarks into three fermion generations defined in Section 14.2. The transitions between the quarks of the same generations, that is, $d \to u$, $s \to c$, and $b \to t$, have the highest rate. The transitions between the quarks of the neighbor generations, such as $s \to u$ or $b \to c$, are less probable. The rate of transitions between the quarks of the first and third generations ($d \to t$ and $b \to u$) is so small that the area of the corresponding squares is not visible at all, and the squares look like dots. Obviously, such a pattern is very peculiar. It currently lacks theoretical explanation, and its understanding represents one of the challenges faced by particle physics.

Thus, the measured rates of quark transitions vary a lot, contrary to naive expectations. So, theorists have no choice but to multiply the expected rates by additional coefficients, which make the theoretical estimates equal to the experimental results. None of the current theoretical models can explain the values of the coefficients introduced.

---

[3]The experiment described has, in fact, never been performed. However, scientists have no doubt that the stated result will be obtained if such an experiment ever take place.

Nonetheless, these correction factors are essential for the following discussion and, therefore, deserve special consideration.

As has already been explained (see Section 9.2), the interaction with a W boson always changes the charge of a fermion by one unit. In particular, this interaction transforms a quark with charge $-1/3$ into one of the quarks with charge $+2/3$ or vice versa. To distinguish these two groups, the quarks with charge $-1/3$ are called *down-type quarks*, and the quarks with charge $+2/3$ are referred to as *up-type quarks*. The charges of all quarks are given in Table 4.1. The down type includes down, strange, and bottom quarks. The up-type quarks are up, charm, and top. The coefficient that adjusts the theoretical value of the transition of a down-type quark x into an up-type quark y is denoted as V_{yx}. For example, the coefficient corresponding to the transition $d \rightarrow u$ is V_{ud}, to the transition $s \rightarrow c$

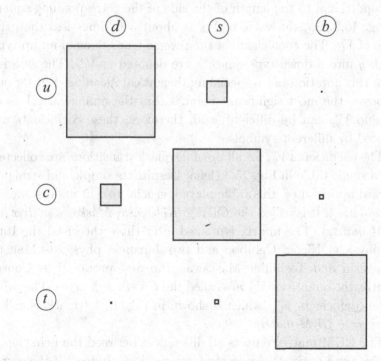

Fig. 15.2 Comparison of the rates of transitions between different quarks. The area of each square is proportional to the rate of the corresponding transition. The largest squares correspond to the transitions $d \rightarrow u$, $s \rightarrow c$, and $b \rightarrow t$.

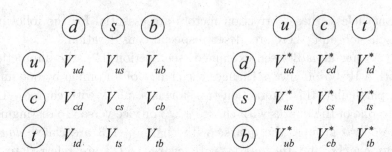

Fig. 15.3   Direct (left) and reverse (right) CKM matrices.

is V_{cs} and so on. Notice that the order of letters (yx) is opposite to the actual direction of the transition $x \rightarrow y$. This inconsistency is due to historical reasons. The absolute value of each V_{yx} coefficient is proportional to the length of the side of the corresponding square in Fig. 15.2. So, the value of V_{ub} is about 300 times less than the value of V_{tb}. The coefficients of the reverse transitions of an up-type quark y into a down-type quark x are denoted as V_{yx}^*. The asterisk (*) in this notation has a special mathematical meaning. But for our purposes, the most significant point is that the numerical values of V_{yx} and V_{yx}^* can be different, and, therefore, these coefficients are denoted by different symbols.

The coefficients V_{yx} for all possible quark transitions are collected into a single table in Fig. 15.3 (left). Despite its simple and straightforward appearance, this table plays a crucial role in understanding CP violation. It is called the *Cabibbo-Kobayashi-Maskawa matrix* (or *CKM matrix*). The matrix is named after three theorists, the Italian physicist Nicola Cabibbo and two Japanese physicists Makoto Kobayashi and Toshihide Maskawa, who first proposed it. Consequently, the quantities V_{yx} are called the *CKM coefficients*. The table of the coefficients V_{yx}^*, which is shown in Fig. 15.3 (right), is called the *reverse CKM matrix*.

The CKM matrix reflects all differences between the behavior of particles and antiparticles in weak interactions. Indeed, if the corresponding coefficients of the direct and reverse CKM matrices were equal, the properties of hadrons and antihadrons would be the same.

Consequently, there would be no CP violation. The opposite statement is also correct. CP violation in weak interactions is possible only if the coefficients of the direct and reverse CKM matrices don't coincide. Hence, the study of CP violation in weak interactions is simplified to the measurement of the coefficients of the direct and reverse CKM matrices and the understanding of why they differ from each other.

15.2   Unitarity Conditions

Although the values taken by V_{yx} cannot yet be justified, they are not completely arbitrary. Special requirements, which are called *unitarity conditions*, relate the CKM coefficients to each other. Scientists first realized the existence of these conditions while attempting to explain some experimentally observed properties of mesons. Later, the unitarity conditions received a proper theoretical interpretation. This route of physics development is very typical. When an unusual experimental effect is detected, theorists try to explain it and derive its possible consequences. Such a study often results in the acquisition of a qualitatively new understanding of nature. The unitarity conditions constraining the coefficients of the CKM matrix serve as an excellent example of this method of scientific research.

The experimental result that initiated the theoretical exploration was the decay of a neutral kaon into a muon and antimuon. In the beginning, the quark theory was unable to correctly describe this decay mode because the model initially contained just three quarks: up, down, and strange. The Feynman diagram involving these quarks is shown in Fig. 15.4 (upper part). At the quark level, the down quark of K^0 meson transforms into an up quark after interaction with the W boson. This transition is proportional to V_{ud}. Subsequently, the up quark interacts with the second W boson and transforms into the strange quark. The coefficient V_{us}^* determines this transition. After all interactions, the down quark and strange antiquark disappear, while the two W bosons, which conduct this annihilation, convert into a pair of oppositely charged muons.

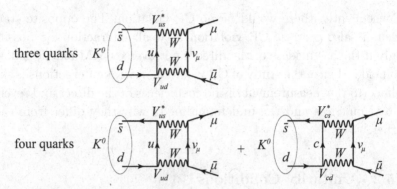

Fig. 15.4   Diagrams for the decay of a neutral kaon into muon and antimuon with three (upper part) and four (lower part) quarks.

The diagram presented allowed for a straightforward calculation of the expected kaon decay rate. However, the obtained result was much higher than experimentally observed. To eliminate this disagreement, the American physicists Sheldon Glashow, John Iliopoulos, and Luciano Maiani hypothesized the existence of a fourth (charm) quark. By adding this quark, they made the second fermion generation (see Fig. 14.1) complete. The contribution of the charm quark to the specified K^0 decay results in an additional diagram shown in Fig. 15.4 (lower part). In this diagram, an intermediate up quark is replaced by a charm quark.

The new diagram changed everything because the theorists realized that the existence of the fourth quark, or more precisely, two fermion generations imposed a special condition on the CKM coefficients. Understanding the need for the condition requires some knowledge of the mathematical language used, but the meaning of the condition can be expressed in simple terms.

The interaction with W boson transforms a down quark into either an up or charm quark. After the second interaction with W boson, the up or charm quarks can change into either a strange or down quark as schematically shown in Fig. 15.5 (left and center). Previously, we said that the "horizontal" transitions, like $d \rightarrow s$, are prohibited in the interactions with a photon or Z boson (see Section 9.2). It turns out that if the masses of all quarks were the same, these transformations would also be forbidden after the interactions

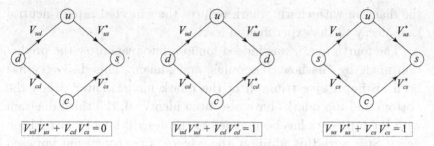

Fig. 15.5 Requirement of the suppression of the transition $d \to s$ in the case of two fermion generations (left). Requirements allowing the recreation of a down quark $(d \to d)$ and strange quark $(s \to s)$ in the case of two fermion generations (center and right, respectively).

with two W bosons. This requirement implies a special relationship between the coefficients V_{ud}, V_{cd}, V_{us}^*, and V_{cs}^*. The relationship is shown in Fig. 15.5 (left).[4] It ensures that the chain of transitions $d \to u \to s$, which is determined by the product $V_{ud}V_{us}^*$, is counterbalanced by the chain $d \to c \to s$, which depends on the product $V_{cd}V_{cs}^*$.

Conversely, the recreation of a down quark from the initial down quark (see Fig. 15.5 (center)) is always allowed. Similarly, the initial strange quark can be "disassembled" into up and charm quarks and, after that, "assembled" again into a strange quark as shown in Fig. 15.5 (right). Technically, all these requirements also imply special relationships between the corresponding CKM coefficients as shown in Fig. 15.5. The number '1' in the right-hand side of the relationships signifies that one initial quark can transform into exactly one final quark.

Because of the requirement shown in Fig. 15.5 (left), the two diagrams in Fig. 15.4 (lower part) almost cancel each other. The net effect would be exactly zero if the masses of an up quark and charm quark were the same. However, the masses are significantly different, so the sum of the two diagrams gives a nonzero result. Still, adding

---

[4]The relationships shown in Fig. 15.5 are given for the case of only two fermion generations. Their extension to three generations is discussed later.

the diagram with charm quark reduces the expected rate of neutral kaon decay to the experimental level.

The fourth quark was indeed found a few years after the prediction made by Glashow, Iliopoulis, and Maiani. This discovery has been an impressive triumph of the quark model. Later, when the bottom and top quarks have also been identified, the third diagram with the top quark has been added to the description of neutral kaon decay. Still, with this addition, the essence of the constraint imposed on the CKM coefficients has remained the same. The overall combination of the CKM coefficients that determine the transformation of a down quark into a strange quark must be equal to zero.

The transition $d \to s$ is not unique, and similar conditions suppress the "horizontal" transformations between other quarks of the same charge. All such conditions for three generations are graphically presented in Fig. 15.6. In each case, the condition states that if the masses of all quarks were the same, the "disassembling" of the down-type quark into three possible up-type quarks and the subsequent collection of the pieces together would not produce another quark flavor.

On the contrary, each quark after the procedures of "disassembling" and "assembling" should be fully restored. This statement is also expressed as the constraints imposed on the CKM coefficients, as shown in Fig. 15.7. The requirements displayed in Figs. 15.6 and 15.7 are collectively called the *unitarity conditions*.

It is important to stress that the unitarity conditions do not completely eliminate the transformation of one down-type quark into

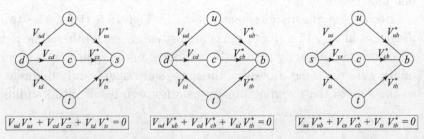

Fig. 15.6   Unitarity conditions for transitions $d \to s$ (left), $d \to b$ (center) and $s \to b$ (right).

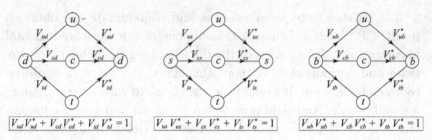

Fig. 15.7 Unitarity conditions for transitions $d \to d$ (left), $s \to s$ (center) and $b \to b$ (right).

another. If the masses of all quarks were equal, this would be the case. However, a significant difference between quark masses results in nonzero residual effects after adding the contributions of all transitions. For this reason, a neutral kaon can decay to a muon and antimuon, although the probability of this process is tiny. Another example of such a "horizontal" quark transition is the oscillation of neutral kaons shown in Fig. 11.1. Here again, a down quark transforms into a strange quark. Due to the unitarity conditions, this transformation is suppressed but is still nonzero.

In total, there are 18 CKM coefficients. However, the unitarity conditions do not allow them to take arbitrary values. We can choose a much smaller set of *free parameters* and unambiguously derive all CKM coefficients from them using unitarity conditions. According to a dedicated study, just four free parameters are sufficient to do the task. But the most interesting result of the study is the following. **Only one out of four free parameters generates the difference between the direct and reverse CKM matrices.** This parameter is called the *CKM phase*. If the CKM phase were zero, the corresponding coefficients of the direct and reverse CKM matrices would be equal to each other. That is, V_{ud} would be equal to V_{ud}^*, V_{td} would be the same as V_{td}^* and so on. Three other free parameters would not change this equality. In this case, the properties of hadrons would not differ from antihadrons, and CP symmetry would be conserved in weak interactions. Conversely, any nonzero value of the CKM phase will automatically produce CP violation.

Let us stop here for a moment and contemplate the obtained result. CP violation is unquestionably one of the most fundamental phenomena, which signifies distinctions between the behavior of particles and antiparticles. Without this effect, the universe would be completely different. It would be a strange world without stars, galaxies, and people. And right now we have realized that just one parameter, the CKM phase, fully determines CP violation in weak interactions. Just one unremarkable quantity, which nobody had even supposed to exist several dozen years ago, decides the fate of antimatter in the entire universe and determines the whole possibility of our existence. Is not it wonderful that the immense edifice of nature is so finely balanced, and that its structure can depend on a single quantity?

In this respect, there is one more interesting fact related to the CKM matrix. This matrix is obtained by assuming three generations of fundamental fermions. If only one generation existed in nature, that is, if a universe contained only up and down quarks, electrons, and electron neutrinos, the only possible transitions would be $d \rightarrow u$ and $u \rightarrow d$. In this case, the matrix of quark transitions would be elementary. It would contain just one coefficient V_{ud}. The unitarity condition modified from those presented in Fig. 15.7 would require that both V_{ud} and V_{ud}^* are equal to one. Consequently, in such a universe, CP violation would be absent.

One generation of fermions is enough for arranging the atomic structure of matter. However, a hypothetical world with only one generation would not contain any CP violation. Hence, the prolonged existence of atomic matter would be impossible. Once again, we can recall Isidor Rabi's question, "Who ordered that?". By asking this question, he wanted to emphasize that a muon, the first particle of the second generation discovered, seemed to be redundant. Today, even though we still don't know who ordered a muon, we can confidently state that Rabi, like everybody else, would not exist without this particle.

What is more interesting, even two generations are not sufficient to produce CP violation. With two generations of the fundamental fermions, that is, with u, d, c, s quarks and corresponding leptons, the matrix of quark transitions would contain just four coefficients: V_{ud}, V_{us}, V_{cd}, and V_{cs}. The unitarity conditions adjusted to the two-generation case leave a single free parameter, which unambiguously defines all these coefficients. Moreover, the same conditions require that the matrices of direct and reverse quark transitions coincide. In other words, the CKM phase, which produces a difference between the direct and reverse CKM matrices, does not exist in a world with just two fermion generations. Thus, two generations still would not permit Rabi to ask his question. At least three generations of fundamental fermions are required for this. Only then, the matrix of quark transitions will depend on the special parameter which generates CP violation.

Here we once again reveal the greatest perfection of nature's arrangement. Everything is meaningful in this world. Each tiny detail matters. The mean lifetimes of all particles from the second and third generations, except invisible neutrinos, do not exceed a few microseconds, and most of these objects live for much shorter lengths of time. Even when they are created in collider experiments, they are almost immediately destroyed. The fact of their existence or their complete absence in nature may seem insignificant. Indeed, most people don't even suspect their presence. Nonetheless, without the particles of the second and third generation, the world in its present form would not exist. Only the existence of three generations makes CP violation possible.

The Japanese theorists Makoto Kobayashi and Toshihide Maskawa were the first scientists to uncover the significance of the third generation of fermions for the origin of CP violation. They considered the matrix of quark transitions for three generations and proved that its structure admits CP violation. From this study, they concluded that **all differences between direct and reverse quark transitions and consequently all possible CP-violating phenomena are determined by a single quantity, the phase of the matrix of quark transitions**. This theory of generating CP violation is now called the *KM (Kobayashi-Maskawa) mechanism*. The third generation was not yet known when Kobayashi and Maskawa published their work. Thus, they, in fact, predicted the existence of the third generation by stating its need to explain CP violation in weak interactions. It is not surprising that, subsequently, Kobayashi and Maskawa were awarded the Nobel Prize for their research.

The development of the KM mechanism is a remarkable achievement in the explanation of CP violation. However, it would be too simplistic to think that the study of CP violation is now reduced to the measurement of the single CKM phase. First, the KM mechanism is a theoretical model. Like any other theory, it needs experimental confirmation. Hence, one of the main directions of research involves testing the predictions of the theory by Kobayashi and Maskawa. Also, there could be some other sources of CP violation in addition to the KM mechanism. If this is so, they should be identified and understood. Finally, despite stating that the source of CP violation is the CKM phase, the KM mechanism does not explain the origin of this phase and does not predict its value. Indeed, why does the CKM phase have a given value and not another? Without answering this question, humankind's understanding of CP violation will never be complete.

As was mentioned earlier, the answer to one scientific question always uncovers a more challenging puzzle. The study of CP violation is not an exception. The origin of the CKM phase, like all other parameters of the CKM matrix, is yet unknown. However, the information required for the verification of the KM mechanism has already been collected. It includes experimentally measured CP-violating effects and other properties of various hadrons. These measurements, as well as the results of the tests of the KM mechanism, are presented in the next section.

15.3   Experimental Studies of CP Violation

The KM mechanism predicts that the single quantity, the CKM phase, explains all observed CP-violating phenomena. This prediction is unambiguous and experimentally testable. A crucial test, which can confirm or disprove the KM mechanism, involves a verification of the unitarity conditions using the CKM coefficients. The motivation of such a *unitarity test* is the following. The theory of

Kobayashi and Maskawa provides formulae that allow scientists to derive the CKM coefficients from the measured CP-violating effects. If the KM mechanism is valid, the obtained CKM coefficients must satisfy the unitarity conditions. Otherwise, if the conditions were not met, the derived CKM coefficients would not be correct. Hence, the derivation of these coefficients from the CP violation effects would be flawed. The invalidity of the derivation would imply that the KM mechanism is not relevant to CP violation. Consequently, the source of CP violation should be searched for elsewhere.

To perform the unitarity tests, scientists measure the *value of CP violation*, that is, the quantitative expression of the effect. For example, Cronin and Fitch observed CP violation in the decays of a long-living neutral kaon into two oppositely charged pions. Such decays happen in two cases out of one thousand regular decays of this particle. Hence, the ratio $2/1000$ reflects the value of CP violation.

According to the theory of Kobayashi and Maskawa, CP violation emerges whenever the coefficients of the direct and reverse quark transitions are different. Hence, by measuring the value of CP violation, scientists can find the difference between the coefficients of the direct and reverse CKM matrices. For example, the difference between V_{td} and V_{td}^* can be obtained from the measurement of CP violation in the processes caused by the $d \to t$ and $t \to d$ transitions.

In addition to the differences between V_{xy} and V_{xy}^*, physicists measure the absolute values of the CKM coefficients, which are also required for the unitarity tests. These values can be extracted from various measurements, which are not directly related to CP violation. For example, the decay of the D^+ meson presented in Fig. 9.8 is caused by the transition $c \to s$. Hence, by measuring the rate of this decay mode scientists get the value of V_{cs}. Similar decays of other hadrons define the corresponding CKM coefficients. Still, the difference between the coefficients of the direct and reverse CKM matrices can be deduced only from the measurements of CP violation.

The study of CP violation is complicated since, in most cases, the contribution of this effect to physical processes is barely noticeable. Because of this, scientists believed for a long time that CP symmetry was always conserved. The minuscule value of CP violation is not caused by the tiny CKM phase, as it may be supposed. The phase itself is reasonably large. The smallness of the value is explained by the suppression of the transitions between different fermion generations. The level of this suppression is visible in Fig. 15.2, where the area of the square corresponding to the transition $d \to u$ is much greater than that corresponding to $d \to t$.

As was discussed in the previous section, CP violation arises only when the third generation is added into the matrix of quark transitions. Therefore, it is natural to expect that CP-violating effects should be maximal in the processes involving the transitions between the quarks of the first and third generations. And this is indeed the case. More specifically, the transitions $d \to t$, $b \to u$, and the corresponding reverse transitions produce the largest CP violation. But these transitions are strongly suppressed in comparison with more probable transitions $b \to t$ and $d \to u$, as can be seen in Fig. 15.2. Consequently, the resulting value of CP violation is extremely small. It is enhanced only when the specified transitions give the main contribution to a process.

Since the maximum CP violation should manifest itself in the transitions $b \to u$ and $d \to t$, bottom and top quarks are the most appealing objects for the measurement of this phenomenon. A top quark has an extremely short lifetime. This lifetime is so momentary that a top quark has no time to even couple with an antiquark and produce a hadron. It decays much sooner. Also, considerable energy is required to produce this particle because of its enormous mass. Hence, experimental setups required to detect top quarks are very complicated, and the use of top quarks for the study of CP violation is prone to many difficulties. On the contrary, a bottom quark is very convenient for these purposes.

First of all, a b quark couples with various quarks and antiquarks and produces so called *beauty hadrons*, or B hadrons. These particles have a reasonably long lifetime. Of course, compared to any living organism, B hadrons decay almost immediately after their production. But modern instruments can distinguish the moments of their birth and disintegration and, consequently, measure their lifetime.

The long lifetime helps to separate B hadrons from other particles. This property is also extensively used in the studies on CP violation.

The second important feature of B hadrons is the broad range of their decay modes. B-hadron decays are determined by transitions of the bottom quark into either an u or c quark. The transition $b \to u$ is needed for CP-violation studies. Therefore, B-hadron decays involving this transition can be used to measure the difference between V_{ub} and V_{ub}^*. Several other decay modes of B hadron are sensitive to the transition $t \to d$, which is also essential for the measurement of CP-violating effects. Thus, B hadrons provide a wealth of information to test the KM mechanism.

There is one more subtle argument in favor of B hadrons. The CKM coefficients are derived from the measured properties of particles using theoretical formulae. But the formulae have a limited accuracy because the contribution of some effects is very difficult to take into account. These effects are mainly related to the strong interaction between quarks contained in a hadron. It turns out that the strong interaction of low-energetic particles is so complicated that the QCD is incapable of describing it precisely. However, with the increase in particles' energy, theoretical predictions become more accurate. The energy of quarks confined in hadrons is determined by their mass, and the mass of a bottom quark is much higher than the mass of a strange quark. Consequently, the theoretical description of CP-violating effects in B decays is much more reliable than that in neutral kaon decays. Hence, the extracted CKM coefficients are of considerably better accuracy.

Because of all the advantages mentioned, B hadrons are currently the main particles used for CP violation studies. These particles include B^+ (with quark content $\bar{b}u$), B^0 ($\bar{b}d$), and B_s ($\bar{b}s$) mesons. Their antiparticles are denoted as B^-, $\bar{B}^0$, and $\bar{B}_s$. B^0 and B_s mesons have no charge and, therefore, can oscillate like the K^0 meson (see Chapter 11). The diagram of B^0-meson oscillation is shown in Fig. 15.8. In this process, a d quark transforms into a t quark, which later transforms into a b quark. In addition to these transitions, the chains $d \to c \to b$ and $d \to u \to b$ are also possible.

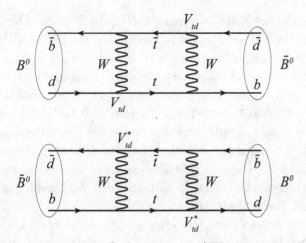

Fig. 15.8   Diagrams of B^0 (upper plot) and $\bar{B}^0$ (lower plot) oscillations.

However, numerical calculations demonstrate that the main contribution to the oscillation comes from the chain $d \to t \to b$.

Since the B^0 oscillation involves the transition $d \to t$, the process depends on V_{td} (see Fig. 15.8). Similarly, the oscillation of the $\bar{B}^0$ meson depends on V_{td}^*. The difference between these two coefficients results in CP violation due to the $d \to t$ transition. Therefore, substantial CP-violating effects can be expected in B^0 decays that are sensitive to B^0 oscillation. That is why studying these decays allows for measurement of the difference between V_{td} and V_{td}^*.

It is interesting to note that CP violation in decays of a long-living neutral kaon (K_L), considered in Chapter 11, is also caused by the difference between the transitions $d \to t$ and $t \to d$. A K_L meson is a mixture of K^0 and $\bar{K}^0$ mesons. This mixture is produced by the K^0 oscillation, which also depends on the $d \to t$ transition (see Fig. 11.2). So, CP violation in K_L decay is determined by the difference between V_{td} and V_{td}^*. However, the theoretical estimate of the contribution of strong interaction into this CP-violating effect is too imprecise because of the low mass of an s quark. Consequently, the difference between V_{td} and V_{td}^* extracted from K_L decays is quite inaccurate. Hence, only oscillating B^0 mesons provide suitable information on the transitions $d \to t$ and $t \to d$.

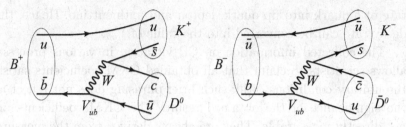

Fig. 15.9 Decay of a B^+ meson to D^0 and K^+ mesons (left diagram). Decay of a B^- meson to $\bar{D}^0$ and K^- mesons (right diagram).

The difference between V_{ub} and V_{ub}^* also produces a considerable CP-violating effect. This type of CP violation can be studied using the decays of B mesons where the transitions $u \to b$ and $b \to u$ give a major contribution. One of such decays is presented in Fig. 15.9. The decaying particle is either a B^+ or a B^- meson. The final state of B^+ decay includes a D^0 meson (quark content $c\bar{u}$) and K^+ meson ($u\bar{s}$). This decay is determined by the transition $u \to b$, which depends on V_{ub}^*. The corresponding decay of the B^- meson, which is also shown in Fig. 15.9, depends on V_{ub}. Hence, scientists extract the difference between V_{ub} and V_{ub}^* by measuring the value of CP violation in these decays.

In addition to the measurements discussed above, CP violation has been observed in many other decays of B hadrons. Each such observation gives an independent measurement of the CKM coefficients, which is included in the unitarity test. Within the KM mechanism, the transitions $b \to u$ and $d \to t$ are the primary sources of CP violation. Other CKM coefficients also differ from the corresponding coefficients of the reverse CKM matrix. However, the expected difference is tiny and has not been experimentally detected yet. Therefore, various observations of CP violation determine only the difference between V_{td} and V_{td}^* and between V_{ub} and V_{ub}^*.

The absolute values of the coefficients V_{td} and V_{ub} are also extracted from B decays. For example, the value of V_{td} is derived from the *oscillation frequency* of a B^0 meson. This quantity defines how quickly a B^0 meson turns into a $\bar{B}^0$. The frequency depends on the value of V_{td} since the transition $d \to t$ is responsible for the oscillation (see Fig. 15.8). The coefficient V_{ub} influences the decay rate of b quark into up quark, lepton and antineutrino. Hence, this decay rate can be converted into the value of V_{ub}.

The collected information on CP violation in various processes allows scientists to claim that all obtained CKM coefficients satisfy the unitary conditions. Still, such brief phrasing does not reflect all the significance of the obtained result. The CKM coefficients are not directly measurable. They are always derived from the measurements of various properties of different hadrons using the formulae predicted by the KM mechanism. This theory is the only connection between otherwise-unrelated effects. Indeed, CP violation in K_L-meson decays looks entirely dissimilar to CP violation in B^+-meson decays. Nonetheless, the KM mechanism successfully merges these and many other unrelated CP-violating phenomena within a single framework. For this unification, the KM mechanism employs just one quantity, the CKM phase. Thus, the successful unitarity tests signify that **the CKM phase is indeed the main source of CP violation**.

The conclusion on the validity of the unitarity conditions has been obtained by the experiments Babar and Belle, which operated at two different electron-positron colliders in the USA and Japan, respectively. The role of these experiments in establishing the KM mechanism is tremendous. It is sufficient to say that Kobayashi and Maskawa were awarded the Nobel Prize only after the two experiments confirmed their theory.

Undoubtedly, the identification of the origin of CP violation is one of the greatest achievements of particle physics. But it is not absolute victory. The unitarity tests cannot exclude the existence of some other, less strong, sources of CP violation. The experimental results indeed agree with the theoretical expectations. However, every experimental result has a limited *accuracy*, which specifies a possible difference between the actual and measured quantities. The validity of the unitarity conditions is also confirmed within certain limits, which depend on the accuracy of measurements. Let us suppose that there is an additional source of CP violation, which is caused by as-yet-unknown phenomena and which is not related to the KM mechanism. If the value of CP violation produced by this source is considerably smaller than the effect due to the CKM phase, and if the accuracy of the experimental measurements is not sufficient, the contribution of such new phenomena will be unnoticeable. At least, it will be impossible to distinguish the additional contribution from the experimental imprecision. Therefore, even though the main source of CP violation is established, nobody can claim that this source is the only reason for the absence of antimatter in the universe.

Let us summarize the results of the study of CP violation presented in this chapter. These results are indeed extensive since particle physics is now able to explain all observed CP-violation phenomena. The explanation consists of the following major points. All transitions between quarks are determined by the coefficients of the CKM matrix. The difference between the coefficients of the direct and reverse CKM matrices generates CP violation. The CKM coefficients are related to each other and must satisfy the unitarity conditions. Because of these conditions, all CKM coefficients are unambiguously determined by just four independent parameters. Only one of these parameters, the phase of CKM matrix, causes all known CP-violation effects. If the CKM phase were zero, CP symmetry would be conserved in the Standard Model. The theory that explains the origin of CP violation by the properties of the CKM matrix is called the KM mechanism. All experimental studies of CP violation agree with the statement that the KM mechanism is the main source of CP violation. However, a possible contribution of additional, less strong sources of CP violation is not excluded. These sources can be revealed by improving the precision of the measurements.

For physicists, the observation of new sources of CP violation would be the most desirable outcome. Above all, scientists always search for something new. This is the main goal of their investigation. But there is also another rationale behind the need for new sources of CP violation. This reasoning, as well as the search for new CP-violating phenomena beyond the KM theory, will be discussed later. However, before proceeding to this part of our story, we need to consider possible implementations of Sakharov's third condition.

Chapter 16

Departure from Thermal Equilibrium

16.1 Thermal Equilibrium in Grand Unification Theories

According to Sakharov's third condition, interactions that generate an excess of matter must be out of thermal equilibrium. This requirement is essential because, in general, for each process that increases the number of particles, a reverse process that reduces this number should exist. In thermal equilibrium, both processes take place. Consequently, any generated excess of matter is washed out by the reverse process, and, with time, the fractions of particles and antiparticles become equal. The third condition aims to prevent this from happening. It stipulates that at the time of baryogenesis, simultaneous action of both direct and reverse processes was prohibited. In this case, an excess of matter should persist.

Meeting the third condition seems easier compared to the first and second ones because, in our normal world, there are many examples of departure from thermal equilibrium. These examples help to design the implementation of the third condition in the microworld of particles. One of such cases, which all of us know, is the *phase transition* of water from a liquid to a gaseous form. Simply speaking, this is the process of boiling. In general, phase transitions of this type are out of thermal equilibrium. Of course, any liquid can become a gas, and a gas can condense into a liquid. In other words, both direct and reverse processes are possible. Drying puddles after the rain, and drops of dew on the grass in the early morning are the convincing examples that evaporation and condensation exist. However, when water boils, the transformation goes on only in one direction. The reason for this is simple. Which of the two processes happens is determined by the flow of energy. We need to heat water to boil it. Similarly, to condense vapor, we must cool it, that is, take away part of its energy. Thus, the requisite of putting in or withdrawing energy breaks thermal equilibrium.

In the microworld, the factor of energy is also vital for departure from thermal equilibrium. Let us again consider the example from Section 12.3. In this example, a Q boson can decay either to a proton and electron, or to an antiproton and positron as shown in Figs. 12.1. We assume that these decays violate CP-symmetry and, hence, generate an excess of matter over antimatter. However, in thermal equilibrium, the reverse transformations of a proton and electron, or an antiproton and positron, into the Q boson wash out any such excess. Now, let us take into account the fact that all these processes happen in the expanding universe. This expansion results in a reduction of the temperature of the universe. As was explained in Section 14.2, the temperature of the universe is equivalent to the energy of radiation that fills the space. When the energy of radiation decreases, the average energy of all other particles also declines. This drop happens because radiation interacts with particles, and energies of both equalize as a result. Such an equalization is quite similar to the cooling of a glass of hot water in a cold room.

Decay of the Q boson always takes place. It does not depend on the expansion of the universe because the energy required for decay is stored in the mass of the Q boson. However, the energy of created particles (a proton and electron) or antiparticles (an antiproton and electron) decreases in the expanding universe. If the expansion is sufficiently rapid and the production of the Q boson is relatively slow, the particles and antiparticles will exist for long enough, and their energies will eventually become smaller than the mass of the Q boson. In that case, Q production will not be possible anymore.

The law of energy conservation will prohibit it.[1] In other words, **rapid cooling of the universe due to its expansion causes the departure from thermal equilibrium**. If the decay of the Q boson to a proton and electron is preferred to the decay to an antiproton and positron, the excess of matter particles produced will stay. Later, the excess will result in a complete disappearance of antimatter.

Such implementation of the departure from thermal equilibrium is pretty realistic. It is effected, for example, in models of grand unification. In all such models, new bosons that violate the baryon number have a huge mass. Consequently, the processes involving these bosons could occur only in an extremely hot universe. The required temperature level was possible only during the first few instants after the Big Bang. Theoretical estimates show that, at that time, the expansion rate of the universe exceeded the rate of processes conducted by heavy bosons of GUT. Because of this, the creation and destruction of baryons by bosons of GUT was out of equilibrium, and therefore Sakharov's third condition was satisfied.

Thus, models of grand unification naturally fulfill the first and third conditions. Nonetheless, all such models are prone to one major difficulty. When the temperature of the universe decreased below the threshold of grand unification, the GUT bosons ceased to mediate interactions and stopped producing new baryons. However, sphaleron processes predicted by the Standard Model (see Section 14.4) also violate the baryon number. Although nobody has observed them yet, they must exist as far as the Standard Model is correct. These sphaleron processes can operate at considerably lower energies compared to the threshold of GUT. Consequently, after the termination of GUT interactions, sphaleron processes continued to create and destroy baryons for a rather long time.

At the epoch of sphaleron processes, the expansion of the universe became moderate. According to theoretical estimates, with the decrease in temperature, the rate of creation and destruction of baryons by the sphaleron processes became *higher* than the universe's

---

[1]The energy of colliding particles cannot be smaller than the mass of the Q boson.

cooling rate. But for the departure from equilibrium, on the contrary, the rate of cooling should exceed the rate of baryon destruction. Hence, the sphaleron processes were in thermal equilibrium. Therefore, they inevitably washed out any excess of baryons generated by GUT. We do not need to perform complicated mathematical calculations to make this statement. It is just sufficient to recall that a physical system always attains the state with equal proportions of baryons and antibaryons as far as both the creation and destruction of them are possible.

Thus, we come to an interesting conclusion. Even if one of the proposed GUT models is ever confirmed, that model will not be the main reason for the antimatter disappearance. GUT models have many attractive features because they combine all interactions and treat them using a uniform approach. However, it turns out that they cannot explain the absence of antimatter.

Looking on the positive side of things, the Standard Model itself could generate an excess of matter in the universe, so that GUT is not needed for this. Both baryon number violation and CP violation are present in the Standard Model. Hence, this theory just needs to demonstrate an implementation of the third condition. In that case, the Standard Model will be able to provide a complete solution for the problem of baryogenesis. Such implementation indeed exists. It is considered in the next section.

16.2   BEH Mechanism

One of the most significant achievements of the Standard Model is an elaborate theory explaining the generation of all particle masses. It is called the *Brout-Englert-Higgs (BEH) mechanism*. This name honors the three theorists, Robert Brout, François Englert, and Peter Higgs, who developed the corresponding theory. According to the BEH mechanism, all leptons and quarks were created massless. They acquired mass sometime after the Big Bang when the temperature of the universe substantially decreased from an initially-huge value. More specifically, the reduction in temperature changed the *vacuum* of the universe, and this change caused particles to get mass. The phrase "change of the vacuum" may seem meaningless. The vacuum is a void, is it not? So, what can be modified there if it contains nothing? But even one single fact, the expansion of space, proves that vacuum is an extremely complicated physical concept. Scientists have already uncovered many of its unusual properties, and what they have revealed is much more astonishing than the most unbelievable fantasies of past dreamers. But this is just the beginning. Lots of new surprising discoveries related to the concept of vacuum are still ahead of us.

According to the BEH mechanism, mass is not an invariable property of particles. The value of particle mass depends on the vacuum which wraps a particle. A particle immersed in the vacuum of one type is massless, while the same particle placed in another type of vacuum acquires mass. We can compare this change to differences between the motion of an object (for example, a metal sphere) in air and water. Evidently, an object always remains the same. However, its comportment in two mediums is dissimilar. In particular, moving in water requires much more effort. Likewise, the behavior of particles differs in the two types of vacuum, and this difference corresponds to the presence or absence of its mass. All particles immersed in one vacuum are massless and, so, move with the speed of light. In the other vacuum, the particles gain mass. Consequently, their speed is less than the speed of light and can change whenever an external force acts upon them. Of course, the actual BEH mechanism is considerably more complicated than this simplistic interpretation. But to appreciate all details of this theory, fluency in the relevant mathematical language is needed. Therefore, a more involved discussion of the BEH mechanism would go beyond the scope of this book.

The practical operation of the BEH mechanism requires the existence of a special particle called the *Higgs boson*. The prediction of this object is a cornerstone of the theory. Scientists searched for the Higgs boson for about fifty years, and this perseverance was finally rewarded. The ATLAS and CMS experiments at the LHC collider discovered the Higgs boson in 2013. All properties of this particle agree well with theoretical expectations. Hence, one more prediction of the Standard Model has been brilliantly confirmed. This fundamental discovery strongly supports the claim of the BEH mechanism that particles acquire mass because of changes in the vacuum, so this hypothesis looks fully realistic.

We can compare two states of the vacuum determining the mass of particles to two phases of water. Within this analogy, the two states of the vacuum can be termed to be *"gaseous"* and *"liquid" phases.* All particles present in the "gaseous" phase of the vacuum are massless, while, in the "liquid" phase, they have mass. Such analogy is quite instructive because the transition of the vacuum from one phase to another has similar features to the phase transitions of water, that is, to boiling and condensation. Both in the case of water and vacuum, their properties are drastically different depending on their phases. Moreover, in both cases, the transition from one phase to another is caused by the change in temperature. For the vacuum, the change in temperature of the universe is what matters. When the temperature was high, the vacuum was in "gaseous" phase. When the temperature decreased, the vacuum transformed into the second, "liquid", phase. Of course, the actual states of the vacuum don't resemble liquid or gas. Nonetheless, the phase transitions of water illustrate well what happens to the vacuum when it transforms from one state into another.

There are two kinds of phase transitions. Boiling and condensation are *first-order phase transitions.* When we heat water, its temperature continuously increases up to 100 degrees Celsius. After that, water starts boiling. During this process, the temperature of the water does not change, and the process of vaporization consumes all incoming energy. Only when all liquid turns into gas, the temperature resumes its rise.

The development of most first-order phase transitions is similar. Within the medium in one phase, small domains of the other phase start to appear. For example, when water starts to boil, small bubbles of gas form inside the liquid. Likewise, condensation begins from small drops that form in the gas. The border between two phases is always clearly distinguishable. With time, the domains that contain the new phase gradually grow, and, finally, all the medium enters the second phase.

The transition from one phase to another can also be of the *second order*, or even a *continuous crossover*. During such transitions, the properties of the medium change steadily. Usually, this change is not even visible. Consequently, the border between two phases cannot be specified.

Just as commonly-seen mediums, the vacuum transforms from one phase into another when the temperature of the universe decreases. This transformation can also be of the first or second order, or a smooth crossover between the phases. The type mainly depends on the mass of the Higgs boson. This dependence is not surprising since the Higgs boson is the central part of the BEH mechanism, and the properties of this particle fully determine the operation of the theory. If the vacuum experiences first-order transition, its properties change discontinuously when the temperature of the universe decreases. If the transition is of the second order or the crossover, the properties of the vacuum change uniformly and gradually.

The similarity between the phase transitions of mediums and that of the vacuum can help one to understand how the Standard Model fulfills Sakharov's third condition. Sphaleron processes play the leading role in this action. The rate of sphaleron processes depends on the phase of the vacuum. In the "gaseous" phase, when the mass of all quarks and leptons is zero, the rate is high, and the creation and destruction of baryons are fast and intense. On the contrary, in the "liquid" phase, sphaleron processes cease to occur. The mass acquired by particles and the lower temperature of the universe are the main reasons for this effect.

Let us consider two cases. First, suppose that the phase transition of a vacuum is of the first order. When the transition begins, small domains of the "liquid" vacuum appear inside the "gaseous" vacuum. Initially, the number of such *bubbles* of new vacuum is rather small. However, it gradually increases, and, more importantly, the size of bubbles rises too. Outside these bubbles, the sphaleron processes are numerous because they act in the "gaseous" phase. These processes violate baryon number conservation and create new baryons and antibaryons. The number of created baryons can be higher due to CP violation. But if the creation of baryons happens far from the bubbles of "liquid" vacuum, the reverse sphaleron processes will have time to eliminate any imbalance between matter and antimatter.

The destinies of particles and antiparticles created near the bubbles of "liquid" vacuum are different since the growing bubbles absorb them. The transition from one domain to another is rapid because of the sharp border between the two regions. Hence, if the excess of particles over antiparticles is produced near the bubbles, the reverse sphaleron processes in the "gaseous" phase do not have enough time to establish thermal equilibrium.[2] For this reason, the excess is transferred inside the bubbles, where the sphaleron processes stop to act. No other interaction of the Standard Model can violate the baryon number. Consequently, the imbalance between matter and antimatter produced in the "gaseous" vacuum will be frozen in the "liquid" vacuum and will remain there forever. Thus, **the first-order phase transition of the vacuum causes the departure from thermal equilibrium**.

Conversely, in the second-order transition, a sharp border between two phases of the vacuum does not exist. Therefore, sphaleron processes don't abruptly halt. Instead, they gradually diminish with the decrease in the universe's temperature. In this case, even if the excess of matter emerges at an intermediate stage, the reverse processes will have enough time to get rid of it. Hence, **the second-order phase transition or the continuous crossover between the phases of the vacuum do not fulfill Sakharov's third condition**.

Let us summarize the obtained results. The sphaleron processes violate the baryon number and, therefore, satisfy Sakharov's first condition. They are in thermal equilibrium for a considerable part of the universe's evolution. Because of this, they destroy any imbalance

---

[2]As specified in Section 16, attaining the state of thermal equilibrium always requires a substantial duration of time.

between matter and antimatter, which can develop immediately after the birth of the universe. However, the sphaleron processes are out of thermal equilibrium during the first-order phase transition of the vacuum. As mentioned already, the order of the phase transition depends on the mass of the Higgs boson. Hence, if the Higgs boson has an appropriate mass, the BEH mechanism will meet the third condition. Finally, the sphaleron processes can also satisfy the second condition because of the KM mechanism, which generates CP violation.

The sphaleron processes, as well as the BEH and KM mechanisms, form part of the Standard Model. Thus, we come to the major conclusion: **the Standard Model contains all necessary ingredients to conform to all Sakharov's conditions**. Consequently, we can try to construct a model of baryogenesis purely based on this theory. Let us now do this.

16.3   Electroweak Baryogenesis

The Standard Model introduces electroweak interaction, which combines electromagnetic and weak forces. That is why its second name is the *electroweak theory*. Consequently, the model of baryogenesis based on the Standard Model is also called electroweak.

The main stages of *electroweak baryogenesis* are the following.

Immediately after birth, the universe was extremely hot. The number of particles and antiparticles created at that time was equal. The vacuum of the universe was in the "gaseous" phase so that all leptons and quarks were massless. The universe swiftly expanded, and its temperature decreased. The sphaleron processes violated the baryon number while CP violation created more particles than antiparticles. However, the sphaleron processes in the "gaseous" vacuum were in thermal equilibrium. Hence, the excess of matter did not survive.

At some moment, the drop in temperature triggered the phase transition of the vacuum into the "liquid" phase. Electroweak baryogenesis assumes that the transition was of the first order, so the border between the "gaseous" and "liquid" phases was sharp. The growing bubbles of the "liquid" phase engulfed the created particles and antiparticles. The transition of particles and antiparticles from one vacuum phase into the other was so rapid that the reverse sphaleron processes had not the time to wash out the excess of matter. Inside the bubbles, the sphaleron processes terminated. Since none of the other interactions within the Standard Model violate the baryon number, the excess of matter over antimatter was frozen.

Finally, all vacuum transformed into the "liquid" phase, and all quarks and leptons obtained mass. By the end of this process, the universe contained an almost equal number of particles and antiparticles. However, the departure from thermal equilibrium during the phase transition of the vacuum resulted in there being approximately one additional particle for every billion particles and antiparticles. After that, myriads of particles and antiparticles annihilated and left nothing except radiation and neutrinos. But the remaining unpaired particles continued to exist. With time, this tiny leftover matter created the universe, which we now observe.

Here, we can again make an intermediate stop and look back over the road that leads to this point. This path is long and arduous. Starting our count from the discovery of the first antiparticle, it took more than eighty years of continuous research for scientists to arrive here. Initially, nothing was known about antimatter. And now, we have a realistic scenario of baryogenesis, which can fully explain the disappearance of antimatter. A notable advantage of this scenario is that it does not need any new theory to work. The well-tested Standard Model is more than sufficient.

Reaching this point in the story of antimatter is, certainly, a great success of science and a giant leap forward. However, we have not reached our destination yet. We have merely passed one more intermediate stage. There are several reasons for this.

First, out of the three essential requirements defined by Sakharov's conditions, only CP violation has received experimental confirmation. The sphaleron processes, as well as the phase transition of the vacuum, have not yet been discovered. The Standard Model, which predicts their existence, has impeccable "credit history". All other predictions of this theory have been successfully confirmed. Therefore, we can reasonably expect that both the sphaleron processes and the phase transition indeed happened at some stage of the universe's evolution. However, the mission to understand our world fully can never be accomplished until scientists experimentally confirm and meticulously study the expected effects.

Second, recently obtained data have revealed a major problem of electroweak baryogenesis. It turns out that **the measured value of the Higgs boson mass prohibits the first-order phase transition of the vacuum**. More specifically, the experimentally obtained value for the mass of the Higgs boson stipulates that the vacuum transition was the smooth crossover from one phase to another. This single problem can ruin the whole beautiful edifice of electroweak baryogenesis, because without first-order phase transition, the departure from thermal equilibrium is impossible in this scheme.

Moreover, all is not well with CP violation as produced by the Standard Model. Within this theory, the KM mechanism is the only source of CP violation. But detailed calculations demonstrate that **this source is not sufficient to create the observed imbalance between matter and antimatter**. The measured excess of matter corresponds to about one particle per one billion particle-antiparticle pairs (see Chapter 13). Nonetheless, even such a small excess seems too big for the Standard Model. The maximum possible effect, which the Standard Model can provide, is about ten billion times smaller. This inconsistency is huge. It means that the universe should contain stars of a number ten billion times smaller than we observe.

Hence, although electroweak baryogenesis can work in principle, its practical implementation does not match the available experimental data. In other words, **the Standard Model alone is incapable of explaining the dominance of matter in the universe**. Some other theory, which eliminates the problems of the Standard Model, is required. Hence, there is still a long path ahead before we can claim to understand the disappearance of antimatter.

The Standard Model is an excellent theory. It superbly explains all observed phenomena of the micro-world. None of the experimental results obtained so far contradict the Standard Model. Even so, the theory miserably fails at the macro-level since it cannot correctly describe the observed structure of the universe. But this failure just means that research must go on. Numerous discoveries of new properties of nature are still awaiting. It means that the best amazements are yet to come. And so, our story of antimatter continues.

The next chapter describes the multiple attempts of scientists to find traces of *new physics*. This term denotes new phenomena which cannot be explained by the Standard Model. The discovery of such effects will help to build a more comprehensive theory, which will correct the shortcomings of the Standard Model and finally explain why there is no place for antimatter in our world.

Chapter 17

Searches for New Physics

17.1 Extensions of the Standard Model

Immediately after the endorsement of the Standard Model as "commander-in-chief" of the microworld of particles, scientists started their attempts to dethrone it. Or rather, they tried to understand whether this theory was indeed the final answer that explains all phenomena related to particles.

The tests of the Standard Model have continued for more than fifty years. During this time, lots of important discoveries have been made. All of them confirm the validity of the Standard Model. For example, the quark structure of hadrons has been established. The tauon and corresponding neutrino, and charm, bottom, and top quarks have been discovered. All these particles are essential components of the Standard Model, and none of them were known at the time of its creation. The W and Z bosons, which mediate weak interactions, have been found. The CKM matrix, which describes quark transformations, has been determined, and its role in generating CP violation has been proven. The gluon, which is responsible for strong interactions, has been identified. A particle, which complies with the predicted properties of the Higgs boson, has recently been detected. Furthermore, many other measurements and tests have not spotted any significant deviation from the expectations of the Standard Model.

This last statement, though, requires two important comments. The first one is related to neutrinos. Initially, the Standard Model treated them as the only massless fundamental fermions. With this assumption, the structure of the Standard Model was especially simple and neat. But the latest studies of neutrinos have proven that these neutral particles have mass like all other fermions. The mass of a neutrino has been found to be at least one million times smaller than the mass of an electron, the lightest charged fermion. In principle, massive neutrinos can be included in the Standard Model without breaking it. However, such light particles are as illogical as the idea of a mammal the size of bacteria. Most probably, a discovery of such an animal would completely ruin Darwin's theory of evolution, or would at least cause its thorough revision. Likewise, a minuscule but nonzero mass of the neutrino deprives the Standard Model of its elegance and creates an unbearable sense of its incompleteness.

Second, with the increase in the number of experiments and improved precision of measurements, the agreement of some of the latest results with the theory is not impeccable. We will consider these results later in more detail. Their interpretation is still ambiguous, so they cannot unquestionably imply that the Standard Model is wrong. However, the appearance of such measurements is notable because it can be the first sign of the advent of new physics.

In any case, an overwhelming portion of the experiments in particle physics agrees well with the Standard Model. On the contrary, several effects detected at the scale of the universe distinctly contradict the theory. The most significant of these observations and the most convincing evidence in favor of the possibility of new physics is dark matter. This form of matter differs from atomic matter, of which the stars, the planets, and all other known objects are made. Particles that compose dark matter should not electromagnetically interact because they do not emit light. They should also be stable and have substantial mass since dark matter contains the largest part of the mass of the universe. Particles with such properties are unknown. At least, there is no place for them in the Standard Model.

Also, as was discussed in the previous chapter, the Standard Model cannot explicate the absence of antimatter in the universe. This failure is especially regrettable because, in principle, the Standard Model contains all ingredients required for the explanation. However, electroweak baryogenesis, which is based on these components, does not stand the tests of experiment.

Hence, a new theory that would extend and complement the Standard Model is not only desirable but also necessary for understanding all phenomena observed on a scale of both the micro- and macro-worlds. Essentially, all scientists agree on this need. But the consensus on the actual realization of the theory is missing. During the long years of research, dozens, if not hundreds, of theoretical models were proposed as the candidates for the title of new theory. The number of such models, which increases each year, is probably limited only by the number of active theorists. All proposed models have the same general feature. They almost exactly reproduce the behavior of the Standard Model at low energies of interacting particles. It is not surprising because the Standard Model nicely explains all known phenomena studied so far. Hence, a deviation from the Standard Model would mean a contradiction of the experimental facts. Only the increase in energy beyond the current limits can reveal any divergence from the Standard Model. That is why experiments at colliders, which are conducted at the highest possible energies, are vital for proving the existence of new physics and for selecting the only correct new theory among multiple candidates.

Many new theories successfully overcome the shortcomings of the Standard Model in explaining the absence of antimatter. One of them is called "Two Higgs Doublets model", or 2HDM. As its name suggests, this theory extends the Standard Model by increasing the number of Higgs bosons. Instead of one such particle in the Standard Model, the 2HDM proposes four. Because of such minimal extension, all interactions of particles that do not involve the Higgs boson are identical to the Standard Model. Any difference between the two theories arises only when the contribution of the new particles is substantial. One of the particles introduced corresponds to the Higgs boson already found, and its mass should be equal to the experimental value. The three other particles must be heavier because only the high mass can justify the absence of contribution of the new particles in the known experimental effects.

The high masses of the three new particles change the operation of the BEH mechanism. The phase transition of the vacuum becomes of first order regardless of the actual mass of the Higgs boson. In this respect, the 2HDM substantially improves the Standard Model, which is incapable of implementing first-order phase transition. Another drawback of the Standard Model consists of an insufficient amount of CP violation generated by the KM mechanism. The 2HDM successfully remedies this weakness too. The supplementary Higgs bosons and interactions introduced by them give rise to additional CP violation absent in the Standard Model. Theoretical calculations demonstrate that this extra contribution is sufficient for explaining the observed abundance of matter in the universe.

Notwithstanding the success in expounding the absence of antimatter, the 2HDM does not contain any new particles that would compose dark matter. Such particles must be stable while the new Higgs bosons of the 2HDM decay very quickly. Hence, this theory, like the Standard Model, does not clarify the nature of dark matter and cannot be the final answer. At best, it can be considered an intermediate stage in constructing a comprehensive new theory.

Other models are much more successful in dealing with dark matter. One such example is the Minimal Supersymmetric Model, or MSSM. This theory introduces a swarm of new particles. For each fundamental fermion, that is, for each lepton and quark, it stipulates the existence of a new boson. Similarly, each boson of the Standard Model, including a photon and a gluon, is paired with a new fermion. Hence, a special kind of symmetry, which is called *supersymmetry*, emerges between all fundamental particles. As always, symmetry confers a wonderful beauty and grace on the model. But this is not the only reason why the MSSM appeals to many physicists. The theory has many other attractive features. Some of the new particles introduced by the MSSM must be stable and electrically neutral. Besides, they need to have a high mass. Such particles are excellent candidates for the constituents of dark matter. Also, with some tuning of the internal parameters, the MSSM overcomes the difficulty of the Standard Model in explaining the disappearance of antimatter.

The imagination of theorists does not end at the 2HDM or MSSM. We mention these models only as examples of a possible extension of the Standard Model. All such improvements have distinct features, which attract more or less numerous groups of scientists. The favorite in this contest is, probably, the MSSM because its team of proponents is especially big and includes both theoretical and experimental physicists. However, confirmation of the validity of this or any other theory is not determined by the number of supporters. It can come only from experimental tests. Fortunately, modern facilities, for example, the LHC collider, allow such verification. The predictions of the MSSM are notably extensive, and the tests of this model can be simultaneously tried in many areas. On the other hand, the number of physicists participating in the LHC experiments is in the thousands. Hence, the possibility of research in hundreds of directions at once is totally real.

Among many feasible targets aimed at the LHC, we highlight the most important ones. Above all, the predictions of properties of the Higgs boson already found differ between the Standard Model and the new theories. Therefore, by measuring these properties scientists can differentiate between various models. The key features, which can help to distinguish the correct underlying theory from other proposed models, are the production rate of the Higgs boson and its decay rate to different final states, such as to pairs consisting of W or Z bosons, b and $\bar{b}$ quarks, or two photons. The ATLAS and CMS experiments measure these rates. So far, all results agree well with the expectations of the Standard Model, but the precision of the measurements is not great. Hence, a possible contribution of new theories could be overshadowed by the usual fluctuations of obtained values due to this imprecision. Therefore, to test the validity of new theories, the experimental precision must be improved. To this end, much higher *statistics* (the number of detected events) are required. Here, the well-known rule that the precision of any measurement is improved with an increase in the collected statistics, applies. This rule is fulfilled not only in physics, but also, for example, in opinion polls. The predictions based on a larger sample of surveyed people are usually more accurate.[1] The experiments at LHC will continue for at least another ten years. During this time, the number of registered decays of the Higgs boson will increase by hundreds of times. Because of this, the precision of the measurements involving this particle will be considerably improved, and the possibility of detecting new physics will become tangible.

Of course, the discovery of unknown particles that have been predicted by a new model would be the most convincing evidence of the model's validity. The search for such particles also goes on at the LHC. So far, all attempts to identify new objects not covered by the Standard Model have failed. Nonetheless, research in this direction continues with great enthusiasm and attracts a significant number of physicists.

Contrary to the measurement of the properties of the Higgs boson, we will not have long to wait for the answer on whether new particles can be discovered at the LHC. Essentially all predicted particles have high mass, and the energy of collisions limits the possibility of their observation. If the mass of new particles exceeds the accessible energy at the LHC, their production at this collider will be impossible. The energy of LHC will not be substantially increased any more. Therefore, if new objects will not be discovered in the nearest few years, they will never be found at the LHC. In that case, more energetic colliders will be required to continue the hunt for unknown particles.

The observation of new objects is particularly critical for the MSSM. This model postulates the existence of many new particles, and none of them have yet been observed. Theorists can explain why

---

[1]The human relations are subject to many other rules and constraints, which can violate the laws of statistics. The notable examples of wrong predictions of the opinion polls are the "Brexit" vote in Britain and the election of Donald Trump as the President of the United States. Nonetheless, physical phenomena strictly comply with all statistical laws.

these particles escape detection, but for this, they need a special tuning of parameters of the model. The number of free parameters in the MSSM is enormous, and theorists succeed so far in their task. However, the tuning of parameters cannot continue indefinitely. At least, some of the new particles must be relatively light and should be observable at the LHC. If this is not so, the coherence and appeal of the MSSM will be severely damaged.

Besides, if the new particles of the MSSM are so massive that they cannot be produced at the LHC, the possibility of explaining baryogenesis using the MSSM will become tricky. The absence of any traces of the MSSM in the experimental data severely limits the freedom in the choice of its parameters. This limitation becomes so restrictive that only a very special and quite unnatural parameter tuning still permits the first-order phase transition of the vacuum. Of course, additional improvements of the model, which consist of adding more new particles, can alleviate this problem. However, there is no other motivation for such an extension except the need to bring the theory into agreement with experiment.

Nonetheless, history knows several examples when a similar tactic of theory adjustment has finally brought success. In 1492, Christopher Columbus set off on a naval expedition from the port of Palos del la Frontera in Spain. He aimed to test a particular theoretical model. At that time, the concept of the spherical Earth was already widely accepted, but its experimental proof was still missing. Because of this, the numerical parameters of the model were misleading, and the expected size of the Earth was much smaller than the actual dimension of our planet. Based on the model with these wrong parameters, Columbus envisaged a shortening of the path to the Orient by moving westward. And so, he started the experiment.

With time, the distance from Europe increased while the shores of the Orient did not surface. Consequently, the parameters of the model were continuously adjusted. To avoid any agitation of his crew, Columbus presented them much smaller estimates of the covered distance and kept the correct numbers only in his personal logbook. Of course, Columbus could declare the failure of the theory somewhere in the middle of the way and turn back. But he did not do this.

Finally, the perseverance of Columbus and his commitment to the set goal bore fruit. He did not discover the western route to the Orient. Instead, he found the whole new world.

The story of Columbus is very illuminating. Above all, it demonstrates that perseverance is an important quality of scientists. New physics may be considerably different from the existing theoretical models, but it will be no less beautiful and wonderful when found. So, the attempts to find new physics must not end regardless of all previous negative results. Just one success will be enough for a complete victory. Will this victory happen? There is only one recipe for revealing this. The show must go on.[2]

17.2   Deviations from the Standard Model

Even if unusual particles predicted by new theories are not found at the LHC, their existence can manifest itself in the deviation of measured properties of particles from the expectations of the Standard Model. The diagram in Fig. 15.7, which shows the oscillation of the B^0 meson, can clarify why such deviations can happen. The mass of the B^0 meson is much lower than the masses of a top quark and W boson, which are involved in the process displayed. Still, the B^0 oscillation takes place. This fact is one of the many miracles of the microworld. The creation of real particles requires adequate energy that exceeds the mass of the objects produced. However, there is no energy constraint for any massive particle to appear in an interaction for a brief time and disappear again after that. Such appearance is called *virtual*. In particular, *virtual loops* similar to that shown in Fig 15.7 can contain particles of any mass. None of the heavy particles can remain in the final state, but their momentary existence in the loop is always possible.[3] In a sense, virtual loops can be called

---

[2]This is the title of the song by the rock group *Queen* (1991).

[3]The effect of virtual particles and virtual loops is explained by the uncertainty principle formulated by the German physicist Werner Heisenberg. This principle states that a particle with energy below its rest energy can exist for a brief period of time. The larger the deviation from the rest energy, the shorter the period at which the particle can exist.

virtual accelerators, and, contrary to ordinary accelerators, the possibility for them to generate new particles is not limited by the energy requirement.

If new particles and interactions exist, they will produce additional virtual loops and contribute to the corresponding processes. The contribution can be substantial even if the mass of the new particles is very high. Because of this contribution, the measurement of particle properties can differ from the prediction by the Standard Model. So, scientists can "smell" new physics without explicitly observing it. In the past, this "smoke" helped scientists many times in selecting the right direction for their research. We can recall the example of neutral kaon decay to positive and negative muons, which was discussed earlier (see Section 15.2). Initially, the experimental data contradicted the simple quark model, which contained only three quark flavors. To eliminate this disagreement, theorists added the additional loop diagram with the then-unknown charm quark (see Fig. 15.4). The mass of a charm quark is much higher than the mass of a neutral kaon. Nevertheless, it can act in a virtual loop and provide the necessary correction to the theoretical expectation. The extension of the theory was very successful because the charm quark was indeed discovered later. Such development of science can repeat itself. That is why another key direction of research at the LHC consists of precise measurement of particle properties and comparison of results with the Standard Model's predictions. Though, there is a caveat to this type of research. In most cases, the probable deviation from the Standard Model is tiny. Hence, all such measurements should be highly accurate.

The most attractive object for these precise measurements is the bottom quark. For this quark, the theoretical predictions are accurate enough while the contribution of new physics into virtual loops can be sufficiently high. The selection of events involving the *b* quark is reasonably easy. Besides, beauty hadrons are copiously produced at the electron-positron and hadron colliders. So, in many cases, the available data are sufficient for the precise measurements. Because of all these factors, the chances of observing effects of new physics in processes with *b* quarks are elevated.

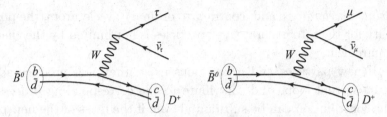

Fig. 17.1   Decay of a $\bar{B}^0$ meson to D^0 meson, τ, and $\bar{\nu}_\tau$ (left diagram). The similar decay with production of μ and $\bar{\nu}_\mu$ (right diagram).

Several measurements of the properties of beauty hadrons indeed demonstrate an intriguing deviation from the Standard Model. We can mention the transition of a b quark into a c quark with the simultaneous production of a tauon (τ) and an antineutrino $(\bar{\nu}_\tau)$. Experimentally, this process is observed as the decay of a $\bar{B}^0$ meson into a D^+ meson, τ, and $\bar{\nu}_\tau$. The corresponding diagram is shown in Fig. 17.1 (left). As always, the strong interaction between the quarks in B and D mesons spoils the precise theoretical estimate of the decay rate. To remove this obstacle, scientists register in parallel a similar B-meson decay that produces a muon and muon antineutrino. The corresponding diagram is shown in Fig. 17.1 (right). The strong interaction between quarks does not depend on the flavor of the produced lepton. Therefore, the ratio of the decay rates of the two processes shown in Fig. 17.1 is accurately calculated in the Standard Model. So, comparison of the theoretical value with the experimental measurement of the ratio represents a powerful test of the Standard Model.

New theories, such as 2HDM or MSSM, can modify the rates of the specified processes. A tauon is much heavier than a muon and an electron, and the influence of new physics can be higher for more massive particles. This expectation is motivated by the BEH mechanism, which states that heavier particles have stronger interaction with the Higgs boson. Something similar may happen in new theories. Hence, the contribution of new physics can amplify the decay that includes a tauon, in comparison to the decay involving a muon. The additional contribution will cause the measured ratio of the two decay rates to deviate from the theoretical expectation. Consequently, the detection of this deviation will be an unmistakable evidence of new physics.

The measurement of the B-meson decay to a D meson, τ and $\bar{\nu}_\tau$ was started in the Babar and Belle experiments, which were already introduced in Section 15.3. The results of both experiments indeed demonstrated a possible disagreement with the Standard Model. However, the statistics collected were too low to claim the contribution of new physics.

The study has been continued by the LHCb experiment at the LHC. The letter 'b' in its name emphasizes that the primary goal of the LHCb experiment is the exploration of beauty hadrons. However, its range of interests is not limited to the b quark. The LHCb physicists have obtained many other impressive results by studying other particles of the Standard Model. Possibly, only study of the top quark is beyond the reach of this experiment.

The first LHCb results agree with that of Babar and Belle and disagree with the Standard Model. The measured ratio of the two decay rates indeed deviates from the theoretical prediction. However, it is too early to claim a discovery. For such a claim, the precision of the measurement should be further improved. Fortunately, the scientists of LHCb expect to collect much more data during the subsequent running of the LHC. Such a perspective is encouraging, as it means that there are excellent prospects for identifying the contribution of new physics.

Proving the deviation from the Standard Model would be a significant achievement. However, the deviation itself gives too little information on the structure of a new theory. The number of possible theoretical models is large. Furthermore, the parameters that determine the operation of the models are not known yet. Hence, predictions of various theories which improve the Standard Model are of poor precision and do not differ too much from one another. Because of this, just the detection of any deviation will not give preference to a particular theory. Only the discovery of new particles not included in the Standard Model, and thorough study of their properties will give an unambiguous answer. Consequently, the search for new particles must continue even if it requires the construction of new colliders.

Besides introducing unknown particles, a prospective new theory must also provide an additional source of CP violation. CP violation generated by the Standard Model is not sufficient to explain the observed abundance of matter in the universe, and the new theory must correct this issue. The additional contribution to CP violation can be detected by comparing the measured value of CP violation with the Standard Model's expectation. Some processes are especially useful for this kind of study. In these processes, the expected Standard Model value of CP violation is close to zero, while the contribution of new physics may be substantial. A large expected value of CP violation is usually subject to sizeable theoretical uncertainties while the prediction of an almost zero value is always accurate. Hence, the non-zero measured value of CP violation in the specified processes will be a clear sign of new physics.

Within the KM mechanism, CP violation is sizeable only in the quark transitions between the first and third fermion generations. In all other cases, its value is expected to be minuscule. This value is not exactly zero because many other coefficients of the direct and reverse CKM matrices slightly differ from one another. However, it is so insignificant that the observation of CP violation in processes that do not involve the transition between the first and third generations is not expected in the Standard Model. Hence, such processes are an ideal place for the search for new sources of CP violation.

In this respect, the transitions between the quarks of the second and third generations are particularly promising. These transitions can also be studied using beauty hadrons. For example, the transition $s \to t$ determines the oscillation of a B_s meson, which contains a strange quark and bottom antiquark. Like in all other cases, this oscillation implies the transformation of a B_s meson into a $\bar{B}_s$ meson. The diagram of this transformation is shown in Fig. 17.2. By inspecting this diagram, we can conclude that CP violation in the oscillation of B_s meson is determined by the difference between the CKM coefficient V_{ts} and V_{ts}^*. In the Standard Model, this difference is tiny. However, new physics can contribute to the virtual loop, which

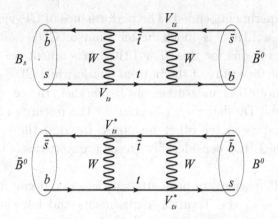

Fig. 17.2   Diagrams of B_s (upper plot) and $\bar{B}_s$ (lower plot) oscillations.

causes the oscillation. Because of this contribution, new sources of CP violation can change the small Standard Model value.

The diagram of B_s-meson oscillation is similar to the corresponding diagram for B^0 meson shown in Fig. 15.7. The only difference between them consists of replacing a d quark by an s quark. Therefore, possible manifestations of CP violation in decays of B_s mesons are similar to those in B^0-meson decays discussed earlier (see Section 15.3). Hence, this type of CP violation should be searched for in the final states where the contribution of B_s oscillation is significant.

The first study of CP violation in B_s decays was performed by the CDF and D0 experiments at the Tevatron collider. This collider operated in the scientific center Fermilab in the USA. The first results of CDF and D0 were promising. Both experiments observed a non-zero value of CP violation in decays of B_s meson, contrary to the Standard Model's expectation. However, the precision of their measurements, which was determined by the available statistics, was insufficient to claim the effect. Moreover, the experimental uncertainties were so large that the interpretation of the Tevatron results also admitted values that were in agreement with the Standard Model.

The ambiguity has been resolved by the LHCb experiment, which has become the leader in studies involving the b quark after the previous experiments ended. The performance of CP-violation measurements in LHCb is much better than in CDF and D0. There are several reasons for this. The LHCb experiment was specifically designed for the study of the bottom quark, while CDF and D0 performed many other measurements in parallel. Hence, the parts of the CDF and D0 detectors essential for the research of B hadrons had been adjusted for other purposes. Besides, the statistics collected by the LHCb considerably exceed the statistics of the Tevatron experiments.

The LHCb scientists have studied the same decay modes of the B_s meson as in the Tevatron experiments and have not observed any deviation from the Standard Model. Meantime, the precision of the LHCb measurement is much better. Hence, the possibility of existence of the new sources of CP violation in the B_s-meson decays considered is not confirmed. All other measurements of CP violation by the LHCb experiment have also not revealed any contribution of new physics. Of course, it does not mean that the treatment of CP violation in the Standard Model is absolutely correct. It just signifies that possible deviations in the value of CP violation are less than the achieved precision of measurements. The LHCb experiment will continue to work for many more years. During this time, the collected statistics will be increased manifold. Hence, the hope of detecting the contribution of new physics to CP violation remains.

17.3   Non-accelerator Searches

Besides the collider experiments, other methods of searching for new physics are also possible. They do not need the collisions of particles. These studies measure properties of stable particles, such as electrons and protons. These particles are everywhere around us, so accelerators are not required to produce them. Meanwhile, the particle properties can be sensitive to the contribution of new physics. The most suitable properties for this kind of research are the *magnetic moment* and *electric dipole moment,* or *EDM.* To understand the meaning of these quantities, we need to descend from the heights of high energy physics and consider low-energetic interactions of charged particles and photons. At low energies, these interactions can be separated into two constituent parts, electric and magnetic. Historically, these two components were first studied separately, and only after that were combined into one, the electromagnetic interaction.

Electric and magnetic phenomena are always conducted by photon exchange, and each such exchange is a discrete, or quantum, process. However, when the number of photons involved is very high, we can describe these phenomena as continuous effects and not bother about quanta. For example, let us consider a negatively charged metal sphere. The charge of the sphere is produced by an excess of electrons in it. The number of extra electrons is enormous, and all of them emit a multitude of photons. The macroscopic action of all these photons upon another charged object, which is placed at some distance from the sphere, manifests as an attractive or repulsive force. According to the classical description of this effect, the charged sphere creates an *electric field*. This entity exists everywhere in the space regardless of the distance to the sphere. The electric field transfers the action of the sphere upon other charged objects and determines the magnitude and direction of the electric force. In this respect, the electric field is like photons. However, the electric field is a continuous thing. It is related to photons the way a stream of water is related to separate drops.

The *magnetic field* produced by magnets is introduced similarly to the electric field. That is, the magnetic field is generated everywhere in space and transfers the action from the source of the field to other objects. However, the electric and magnetic fields are considerably different. The electric field acts upon any charged object, and the force produced is directed along the direction of the field. In contrast, the magnetic force acts only upon moving charged objects. Furthermore, it is always perpendicular to both the velocity of a moving object and the direction of the field.

A spinning classical charged object is subject to the action of magnetic field even if it does not move. It happens because any classical object always has dimensions, and the charge is distributed across its volume. When the object spins, the charges rotate around the rotation axis, and the magnetic field, as always, acts on these moving charges. Because of this action, the object gets additional energy, which is proportional to the magnitude of the magnetic field.

The coefficient that relates the extra energy to the magnetic field is called the *magnetic moment* of the spinning object.

The fundamental particles of the microworld, such as an electron or a quark, have a property similar to the rotation of classical objects. This property is also called the *spin*. The spin of particles is a *vector* quantity, that is, it has a magnitude and direction. The analogy between the spin of classical objects and that of fundamental particles is not entirely correct because fundamental particles are point-like objects and cannot spin in the usual sense. Nevertheless, this analogy is useful for understanding the concept of particle spin. For example, the direction of particle spin is like the direction of the rotation axis of a classical object.

Because of their spin, fundamental particles placed in a magnetic field also get additional energy. Like in the classical case, this energy is proportional to the product of the magnitudes of the spin and the magnetic field. The coefficient of proportionality is also called the *magnetic moment* of a particle. The magnetic moment of all particles of the same type is identical. In many cases, its value can be theoretically predicted with exceptionally high precision. Besides, the magnetic moment can be accurately measured and thus, compared with the theoretical expectation.

The high precision of the theoretical expectation comes with a price. Attaining it requires an enormous volume of calculations. For example, a group of scientists led by the Japanese-American theorist Toichiro Kinoshita spent more than twelve years calculating the magnetic moment of the electron. Of course, they used a powerful supercomputer, but their work still took so much time. Nevertheless, the obtained result was arguably worth these efforts. The theorists have finally reached a precision better than one part per billion. For comparison, to achieve comparable accuracy, we would need to measure the distance between London and Paris with a precision greater than 0.5 mm.

The calculation of the magnetic moment is performed within the framework of the Standard Model. However, the contribution of new physics, as always, can change the actual value. Hence, the measurement of the magnetic moment with precision comparable to the calculations can reveal the existence of new phenomena beyond that specified by the Standard Model. That is why the property of magnetic moment is very attractive for scientists.

In this kind of research, the most promising particle is not an electron, but a muon. The reason for this is the possible sensitivity of new theories to the mass of particles. The mass of a muon is more than two hundred times larger than the mass of an electron. Therefore, the possible contribution of new physics can be greatly increased in the muon's magnetic moment making it more observable. The group of Kinoshita has also calculated the expected value of the muon's magnetic moment and obtained precision greater than one part per million.

Experimentally, the muon's magnetic moment can be measured with an equivalent degree of accuracy. The first precise measurement of this quantity was performed in CERN in 1961, and the latest result so far was obtained in 2006 in the Brookhaven National Laboratory in the USA. During this long stretch of time, the accuracy of the measurement was improved by more than twenty thousand times. Currently, the experimental value of the muon's magnetic moment is known with a precision better than one part per million. Hence, the theoretical and experimental precisions are similar.

The comparison of the expected and measured values of the muon magnetic moment reveals an intriguing discrepancy between them. The disagreement is tiny, in the seventh digit after the decimal point. However, for physicists, it is as substantial as the difference between a cat and a lion. At least, it is sufficient to excite a huge interest in many scientists. After the release of the result, it was discussed in more than four thousand scientific publications, in which theorists attempted to interpret the inconsistency. Of course, the most desirable explanation would be the involvement of new physics. However, as in many other cases, it would be premature to claim this contribution. The experimental and theoretical precisions should be improved by about two times more before such a claim can be made. In any case, the observed discrepancy gives hope that the advent of new phenomena in the microworld of particles is just around the corner.

Another property of particles that can be accurately measured and can testify the involvement of new physics is the *electric dipole moment* or EDM. The property is defined similarly to the magnetic moment, except that the electric field replaces the magnetic one. More specifically, when a particle possessing the EDM is acted upon by an electric field, it gains extra energy proportional to the product of the electric field and the particle spin. The coefficient of proportionality in this relation is the EDM.

Regardless of the similarities in the definition of the magnetic moment and EDM, these quantities are significantly different. Many particles have magnetic moments while none of them have EDMs. There is a serious reason for this division. Any fundamental particle with a non-zero value of the EDM would cause the violation of T-symmetry. To understand this effect, we can consider the action of T-conjugation, discussed in Section 10.3, on a positively charged particle with EDM. When such a particle is placed in an electric field, it moves with acceleration because the electric force acts upon it. When the time is reversed, the particle moves backward and decelerates — see Fig. 17.3. This behavior does not differ from the process in the normal flow of time forwards, in which the electric field has the same direction while the particle has an opposite velocity and, therefore, is decelerated by the electric field. Hence, T-conjugation reverses the direction of the particle's motion but does not change the direction of the electric field.

Let us now consider the action of T-conjugation on the particle's spin. For a classical object, the time reversal changes the direction of its rotation — see Fig. 17.4. For example, the Earth rotates counterclockwise. If the time is reversed, the Earth's rotation will be clockwise. This change in the direction of rotation is equivalent to a reversal of the direction of the rotation axis. That is, after T-conjugation, the rotation axis of the Earth will be directed from the North pole to the South. Likewise, the direction of the spin of elementary particles will be reversed after T-conjugation.

Hence, T-conjugation reverses the direction of spin and does not change the direction of the electric field. These two facts lead to an interesting conclusion. A particle with EDM that is placed in the

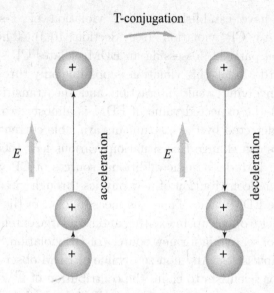

Fig. 17.3   Action of time reversal on a particle moving in an electric field.

T-conjugation

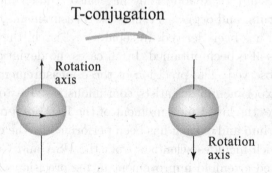

Fig. 17.4   Action of time reversal on a rotating particle.

electric field has an additional energy which is proportional to the product of the electric field and the spin. Hence, after T-conjugation, the sign of this additional energy changes. Consequently, the total energy, which includes the additional contribution due to the EDM, becomes different. In other words, the properties of the physical system before and after the time reversal are different, and this variation corresponds to the violation of T-symmetry.

As we have established earlier, violation of T-symmetry is equivalent to CP violation (see Section 10.3). Therefore, any fundamental particle possessing an EDM violates CP-symmetry. In the Standard Model, this violation is possible only through the loop diagrams involving weak interaction and the transitions between quarks, but the expected value of EDM is almost zero. At least, it cannot be detected by the instruments available currently. However, new physics can change this situation. Various new theories beyond the Standard Model assume additional sources of CP violation. So, the additional contribution of new physics through virtual loops can increase the EDM value. Thus, the measurement of the EDM of different particles opens up an exciting and, to a large extent, surprising, possibility of searching for new sources of CP violation, without producing antiparticles. Any non-zero value of EDM observed would be sufficient for scientists to claim the contribution of new physics.

Several precise measurements of EDM have already been performed. Scientists have measured the EDM of various atoms, such as thallium, cesium, or xenon. They have also studied molecules containing thorium and oxygen. From these measurements, the EDM of the electron has been derived. In separate research, the EDM of the neutron has also been obtained. In all cases, no deviation from zero has been observed. The precision of these measurements is impressive, and experimental scientists continuously strive to improve it. For example, the latest measurement of the EDM of a molecule containing thorium and oxygen has been performed by the collaboration ACME, which includes scientists from the USA and Canada. They have achieved a tenfold improvement in the precision of the derived EDM of the electron in comparison with previous results. Still, the measured value of EDM is consistently zero.

Nonetheless, the obtained result is highly valuable as it considerably constrains the value of CP violation expected in theories beyond the Standard Model. This constraint casts doubt on whether it is possible for some new models, including the MSSM, to explain baryogenesis.

Let us draw some conclusions. As can be seen, the search for new physics goes on in many directions. The results of this search have not been entirely satisfactory. On the one hand, scientists are confident that the Standard Model is not the ultimate theory. It needs to be extended to explain the observed structure of the universe. On the other hand, none of the new particles, which are copiously predicted by different extensions of the Standard Model, has been found yet. Also, none of the additional sources of CP violation has been identified. All CP-violating phenomena have been correctly described so far by the KM mechanism.

Diverse and extensive studies at both extremely high and super-low energies have spotted several weaknesses in the Standard Model and identified a few possible deviations from its predictions. However, definitive proof of new physics has not been obtained yet. Even if such proofs are found and the violation of the Standard Model is demonstrated, the observed deviations will not reveal too much about the nature of new physics. Only the direct production of new particles can validate any new theory.

Nonetheless, the fact that several possible deviations already have been detected is significant. During the Columbus expedition, when the confidence in success began to fade, the sailors started to notice birds. The birds do not live in the open sea, and their appearance meant just one thing: the ships were approaching a new land. The birds provided no information on the size of the land or on its nature. However, they heralded that the new land was close and that just a little more effort was required to reach it. The observed deviations from the Standard Model possibly signal the same thing. They give hope that the discovery of new physics is imminent.

In any case, the totality of obtained results does not clarify the disappearance of antimatter. Hence, investigation of this disappearance must continue. Up to this point in our story, all directions of research have been related in one way or another to quarks and their transformations. However, besides quarks, the microworld of particles also contains leptons. Their role in baryogenesis has not been discussed yet. Nevertheless, this role can prove decisive. We will present later one more model of baryogenesis, which is based on leptons. But before proceeding to this model, we need to consider, in more detail, the properties of the most enigmatic particle among all known objects of the Standard Model. This particle is a neutrino.

Chapter 18

Neutrinos

18.1 Neutrino Oscillation

All previous attempts to explain the absence of antimatter employed predominantly quarks, which were treated as the leading actors in the drama of baryogenesis. Meanwhile, leptons, the second half of the microworld, were mostly considered to be accompanying particles whose role was secondary and inessential. For example, Sakharov's first condition demands the violation of the baryon number and says nothing about the lepton number. Hence, to confirm this condition, scientists search for proton decay and do not worry about the disappearance of leptons. The quark transitions explain the violation of CP-symmetry in the Standard Model. Therefore, CP violation is studied in decays of hadrons. The role of leptons in this line of research is reduced to the identification of required hadron decays. And finally, the term "baryogenesis" implies the creation of baryons (that is, objects composed of quarks) and speaks nothing of *leptogenesis* (that is, the making of leptons).

However, leptons, like quarks, change their flavor in weak interactions too, and this change could also induce CP violation. For example, the diagram shown in Fig. 17.1 (left) displays two simultaneous transitions: $b \rightarrow c$ and $\nu_\tau \rightarrow \tau$. This observation induces a natural question. To what extent do leptons influence CP violation? In a wider context, we can ask about the role of leptons in baryogenesis. Can it be something more than the tedious task of bit player?

We start answering this question from CP violation. It is true that leptons change their flavor like quarks. However, the most apparent difference between the flavor change of quarks and leptons is that each quark can transform into three other quarks while any lepton transmutes only into another lepton of the same fermion generation. For example, a muon can create only a muon neutrino and never an electron neutrino. This limitation excludes the possibility of CP violation similar to the KM mechanism, which needs transitions between different generations for its operation. However, the restriction of lepton transitions can be a purely superficial effect caused by the specific characteristics of neutrinos.

Neutrinos are stable neutral particles. Hence, we cannot determine the type of neutrinos or any other property through their existence. We can distinguish three types of neutrinos only by their interactions. We know that an electron can transform into some invisible particle and that this particle later produces an electron again. That is why we assume that an electron transmutes into an electron neutrino. Nonetheless, lepton transitions can be much more complicated than they look.

Suppose an electron neutrino (ν_e) is the mixture of three particles, which we will call the *first neutrino* (ν_1), *second neutrino* (ν_2), and *third neutrino* (ν_3). Also, suppose that ν_1, and not ν_e, is united together with an electron in the first generation. Similarly, the second and third generations of fermions contain ν_2 and ν_3, respectively. In this case, the transformation $e \rightarrow \nu_e$ represents, in fact, the combination of three transitions: $e \rightarrow \nu_1$, $e \rightarrow \nu_2$, and $e \rightarrow \nu_3$. Likewise, we can suppose that muons and tauon neutrinos are other mixtures of ν_1, ν_2, and ν_3. In this case, a muon and tauon can transform into neutrinos of three different generations, too. In such a model, the structure of lepton transitions becomes essentially the same as the structure of quark transformations. Leptons switch generations the way a bottom quark can become either a top, charm or up quark. Consequently, a matrix of lepton transitions, similar to the CKM matrix of quark transitions, can be introduced. This matrix should obey the corresponding unitarity conditions. Such an extension of the

Standard Model can also contain additional CP violation induced by an analogue of the KM mechanism.

Hence, the proposed scheme offers an attractive possibility of an extra source of CP violation. However, to validate it, the existence of ν_1, ν_2 and ν_3 as distinct particles must be proven. This task is challenging. Three up-type quark flavors (up, charm and top) are easily distinguishable. First, they have clearly distinctive masses. Second, their decay modes are utterly dissimilar and cannot be confounded. In the case of the neutrino, both these handles are absent. Neutrinos never decay. Moreover, scientists thought for a long time that all neutrinos are massless particles. In such conditions, any difference between ν_1, ν_2, and ν_3 is impossible.

Let us consider the case of massless neutrinos more attentively. Suppose that weak interaction of an electron produces an electron neutrino at a point A, and that the created particle is the mixture of massless ν_1, ν_2, and ν_3. All neutrinos in the mixture move with the same speed of light (because they are massless) and in the same direction. Consequently, they simultaneously attain another point B where they interact. Since all three neutrinos reach the point B at the same time, the composition of the mixture does not change and corresponds to an electron neutrino. An electron neutrino always transforms into an electron. So, whenever a mixture called an electron neutrino is created at the point A, it will always produce an electron at the point B. Thus, if neutrinos are massless, we will never know whether an electron neutrino is a single particle or a combination of three other objects.

If ν_1, ν_2, and ν_3 have distinctive masses, the outcome of neutrino interaction will be entirely different. The most noticeable change compared to the massless case is that the three components of the ν_e mixture will propagate at different speeds.[1] Naively, we could guess that the only consequence of different speeds is a spread in the travel time of ν_1, ν_2, and ν_3 when they move from the point A to the

---

[1]The speed of two particles having the same energy but distinctive masses is different.

point B. However, the actual effect will be much more complicated. To understand what happens to massive neutrinos, we need to take into account one of the most fundamental properties of all objects of the microworld.

When we introduced a photon, we said that light could be treated as both a flow of particles and a propagation of a wave (see Section 4.1). This *wave-particle duality* is characteristic not only of photons. It is an inherent quality of all particles, both fermions and bosons, massive and massless. All of them have properties both of a particle and a wave. Although this unity of opposites sounds odd for such material objects as electrons, it has been experimentally proven multiple times and has even been implemented in many practical applications. For example, the wave-particle duality has been actively employed in the instrument called the *electron microscope*, in which examined objects are "illuminated" by the flow of electrons instead of normal light.

One of the key parameters of any microscope is *resolvable distance* (or resolution), that is, the minimum distance between two points that are seen separated in the image produced. When two points overlap in the image, no further magnification can split them. So, improving the resolvable distance is essential for a better performance of the instrument. The microscope resolution mainly depends on the energy of photons in the beam of light that illuminates objects studied. Higher energy results in a better resolving power. However, generating energetic electromagnetic radiation is quite difficult, and working with it is unsafe when energy becomes too high. In this respect, using electrons is much more convenient. They are charged particles and, therefore, can be easily accelerated by an electric field. Hence, they can gain considerable energy, which can be many thousands of times higher than energy that photons can achieve. Since the flow of electrons possesses the properties of a wave, an image of the examined object is produced when electrons scatter off it. Because of the high energy of electrons, the resolution of such images obtained is thousands of times better than that of typical microscopes, which use light. That is why electron microscopes are widely used in science and industry. Their successful operation decidedly confirms that wave-particle duality is a real property of nature.

The flow of neutrinos can also be treated as a propagation of a wave. Since the waves corresponding to three neutrino flavors move with different speeds, their overlap produces a fascinating effect. The composition of the neutrino mixture changes depending on the distance from the production point A. In other words, the mixture which reaches the point B does not correspond to an electron neutrino anymore. It becomes a combination of electron, muon, and tauon neutrinos.

The commonly-seen soap bubble is a good example of a similar phenomenon in our normal world. The bubble can be presented as a thin water film that forms a hollow sphere. When a ray of light is incident on the bubble, it is reflected from the external and internal surfaces of the sphere. Consequently, two reflected waves of light are produced. These waves interfere with each other, and their *interference* can be either *constructive* or *destructive*. That is, the two waves can either add and amplify the reflected light, or they can partially or totally cancel each other out. The outcome depends on the color of light and the thickness of the film.

Now, let us consider a beam of white light incident on a soap bubble. The beam contains the waves of all colors. However, because of the above effect, the waves corresponding to certain colors are dimmed in the reflected light while others, on the contrary, are intensified. Hence, the reflection of white light changes its composition, and, consequently, changes the color. For the same reason, the surface of a soap bubble looks rainbow-colored.

Something similar happens to neutrinos. The three neutrino waves created at the point A move with different speeds. Because of this difference, the neutrino waves interfere while they propagate to the point B. Due to the interference, the composition of the neutrino beam changes, akin to how the color of light alters after reflection from the soap bubble. The fraction of one neutrino flavor is increased; the fraction of another is reduced. Because of this change, the modified mixture contains the contributions corresponding to muon and tauon neutrinos. Hence, when the new mixture is converted into a charged lepton, a muon and tauon should be produced in addition to an electron. The production rate of "alien" leptons depends on the distance between the points A and B. Initially, the rate increases with distance, but after attaining a maximum, it decreases to zero and continues to oscillate after that. Because of such behavior, the effect is called *neutrino oscillation*. This name is accordant with the oscillation of neutral mesons considered earlier (see Chapter 11). These phenomena indeed have a lot in common. Both are caused by the wave nature of particles, and, hence, give similar results.

When the production of muons and tauons increases, the portion of created electrons must decrease because the total number of neutrinos in the beam does not change; they just convert from one flavor to another. In other words, one initial electron neutrino can produce only one charged lepton. Hence, we should detect fewer electrons in comparison with what can be expected without oscillation. This effect is called the *electron disappearance*.

Both the appearance of muons and tauons, produced by an electron neutrino beam, and the disappearance of electrons, are detectable. The observation of this phenomenon should prove that electron, muon, and tauon neutrinos are the mixtures of three other particles with distinctive masses. In turn, this corroboration should open the possibility for lepton transitions between generations and, consequently, for the new source of CP violation.

In the following section, we consider breakthrough experiments that have indeed established the astonishing phenomenon of neutrino oscillation.

18.2   Detection of Neutrino Oscillation

The most powerful source of neutrinos here on Earth is the Sun. Nuclear reactions that happen in our star emit a vast number of electron neutrinos. The fact that these neutrinos are indeed ν_e and not $\bar{\nu}_e$ or ν_μ is precisely known. It is the only possible outcome of the nuclear reactions in the Sun. Neutrinos move in all directions and, in particular, pass through the Earth. Almost all of them traverse the planet without stopping, but some neutrinos do interact with matter and convert into charged leptons. By detecting these interactions, scientists can measure the flow of neutrinos discharged by the Sun and, in many cases, determine their flavor.

The expected rate of neutrino interactions is predicted with reasonably good precision by a theory which describes nuclear processes in the Sun. As always, this prediction can be compared with experimental results. The definitive study of solar neutrinos was first performed by the *Sudbury Neutrino Observatory*, or *SNO*. In this experiment, scientists used *heavy water* to detect neutrino interactions. The molecule of normal water contains two atoms of hydrogen and one atom of oxygen. Instead of the atoms of hydrogen, the molecule of heavy water includes two atoms of deuterium.[2] Such atomic structure was essential for the study of neutrino oscillation by the SNO.

If neutrino oscillation occurs, part of the electron neutrinos will be converted into muon and tauon neutrinos during the journey from the Sun to the Earth. So, the number of electron neutrinos arriving at the Earth will be reduced. An electron neutrino interacting with a neutron contained in the nucleus of deuterium will produce an electron (see Fig. 18.1) while muon or tauon neutrinos will never do this. Hence, neutrino oscillation will cause a smaller number of detected electrons in comparison with the non-oscillating case.

The production of charged electrons is not the only possible result of neutrino interaction with a nucleus of deuterium. A neutrino can

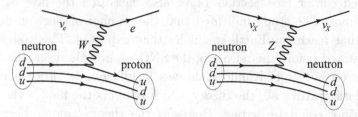

Fig. 18.1 Production of an electron in interaction of an electron neutrino with a neutron (left diagram). Interaction of a neutrino and a neutron mediated by the Z boson (right diagram). The flavor of neutrino is not specified and is denoted as ν_X. That is, it can be ν_e, ν_μ, or ν_τ.

---

[2]The nucleus of deuterium is made of a proton and a neutron.

also kick out a neutron from this nucleus, see Fig. 18.1(right). This process is mediated by the Z boson, which never changes the flavor of particles. Hence, the particles in the initial and final states are the same. However, a neutron after interaction gets additional energy and escapes from the nucleus. Contrary to the first process, both electron, muon, and tauon neutrinos can eject a neutron. That is why the type of neutrino in the diagram on the right in Fig. 18.1 is denoted as ν_X. Therefore, even if a portion of electron neutrinos change flavors while traveling from the Sun, the number of neutrons produced by neutrino interactions will remain the same and will match the theoretical expectation.

Thus, the main idea of the SNO experiment consisted of simultaneous detection of the production of electrons and neutrons. If neutrino oscillation happens, it will reduce the number of identified electrons and will not change the number of neutrons in comparison with the theoretical expectation. By detecting both these effects, scientists could unambiguously prove the validity of the phenomenon of neutrino oscillation.

The SNO experiment was conducted from 1999 to 2006. During this time, researchers obtained a compelling piece of evidence in favor of neutrino oscillation. The number of produced electrons indeed was lower than predicted by the theory. On the contrary, the number of neutrons produced was in perfect agreement with the prediction.

Besides SNO, the Super-Kamiokande experiment, which was considered earlier (see Section 14.3), also measured the flow of solar neutrinos. The study indicated that the number of electron neutrinos that reach the Earth is smaller than expected. This conclusion was stated before the release of the SNO results. However, the effect observed by Super-Kamiokande was insufficient to prove neutrino oscillation. After all, the theory, which predicts the flow of the solar neutrino, could be wrong. However, the theory cannot be wrong in one prediction and be correct in another. Hence, only after the combination of the results of both experiments, and after measuring the number of both electrons and neutrons, the phenomenon of neutrino oscillation was then firmly established. In turn, this has proven the existence of ν_1, ν_2, and ν_3 as distinct particles, which mix into electron, muon, or tauon neutrinos. Subsequently, the leaders of the SNO and Super-Kamiokande experiments, Arthur McDonald and Takaaki Kajita, were awarded the Nobel Prize in 2015 for this outstanding achievement.

Besides the electron neutrino, the muon and tauon neutrinos can oscillate, too. The transitions $\nu_\mu \rightarrow \nu_\tau$ and $\nu_\mu \rightarrow \nu_e$ have already been proven. Using the measured rates of all detected transitions, scientists determine the coefficients that quantify the lepton transformations from one flavor to another. These coefficients are combined into a matrix similar to the CKM matrix introduced in Section 15.1. The matrix of lepton transitions also has the proper name. It is called the *PMNS matrix* to honor Bruno Pontecorvo, Ziro Maki, Masami Nakagawa and Shoichi Sakata, who participated in the development of the theory of neutrino oscillations. It is interesting to mention that, historically, the PMNS matrix for leptons was proposed before the CKM matrix for quarks. Also, Kobayashi and Maskawa, the co-authors of the CKM matrix, were the students and followers of Sakata.

The PMNS and CKM matrices are quite similar. In particular, the PMNS matrix must satisfy the unitarity conditions. Also, differences between the direct and reverse lepton transitions, such as between $e \rightarrow \nu_3$ and $\nu_3 \rightarrow e$, should generate CP violation in lepton interactions. This type of CP violation has not yet been detected. However, its discovery will be a significant achievement in the study of missing antimatter. Therefore, the search for CP violation in neutrino processes is counted among the most essential tasks for current and future experiments.

18.3 Majorana and Dirac Neutrino

The phenomenon of neutrino oscillation is possible only if all neutrino flavors have distinctive masses. If the masses of neutrinos are nonzero but equal, the propagations of the ν_1, ν_2, and ν_3 waves are exactly the same. Consequently, the three waves do not interfere with one another, and the composition of the neutrino beam will not change.

Hence, the detection of neutrino oscillation unequivocally confirms the nonzero neutrino masses.

Despite this remarkable success, the value of the neutrino mass has yet to be determined. It cannot be extracted from neutrino oscillations because they depend only on the mass difference. The possibilities of the measurements of mass by usual techniques are also limited. That is why, currently, only the maximum possible value of the neutrino mass is known. Surprisingly, the best constraint is derived from cosmological studies and not from particle physics experiments. According to these results, all neutrinos must be at least one million times lighter than an electron, the lightest charged particle.

Such a small possible mass worries many scientists. According to the BEH mechanism, the values of the masses of all leptons and quarks are determined by their interaction with the Higgs boson. A stronger interaction results in a higher particle mass. For example, a top quark is about forty times heavier than a bottom quark. Consequently, the top quark's interaction with the Higgs boson is about forty times stronger. The first results of the LHC experiments confirm this expectation and, hence, provide additional confidence in the validity of the BEH mechanism.

The masses of all particles except neutrinos differ by no more than one hundred times within one fermion generation. Such a spread in value seems acceptable, taking differences in particle properties into account. However, the neutrino mass lies far beyond this range. If the neutrino mass were zero, it would be justifiable. It would be interpreted as the absence of neutrino interaction with the Higgs boson. After all, some particles interact with photons; some do not. The same thing can happen to the Higgs interaction. Initially, the Standard Model assumed massless neutrinos, and its structure, with this assumption, was logically sound.

However, neutrinos have a mass, which is millions of times smaller than the masses of other particles. Therefore, the interaction of neutrinos with the Higgs boson must be millions of times weaker in comparison with other particles. Many scientists think that such a colossal gap is impossible. They use this argument to pursue the development of alternative theories that would explain the small neutrino mass. One of these theories is the most attractive because it not only elucidates the small mass of neutrinos but can also help in the question of vanished antimatter.

Neutrinos are truly singular particles. They do not have charge nor color. Hence, they do not interact electromagnetically or strongly. Their weak interactions are also considerably limited. Left-handed neutrinos interact with W and Z bosons (see Section 10.1). However, right-handed neutrinos are deprived even of such possibilities. They do not participate in interactions with a W boson, and it turns out that they do not interact with a Z boson either. In other words, right-handed neutrinos do not take part in any known interaction, except, probably, interaction with the Higgs boson. This remaining string, which attaches right-handed neutrinos to our world, is so thin[3] that it is experimentally undetectable. Thus, from the point of view of electroweak and strong interactions, right-handed neutrinos do not exist.[4] By itself, this fact is amazing. It demonstrates that nature can admit objects that do not display any sign of presence and are, nevertheless, absolutely real.

Another singular property of neutrinos is also related to their zero charge. It turns out that a neutrino and antineutrino can represent the same object. The charges of all particles and antiparticles are opposite. Therefore, charged particles and antiparticles are distinct. However, if a particle is neutral, it can be indistinguishable from its antiparticle. One example of such an object is a photon, or K_S and K_L mesons considered earlier (see Chapter 11). A neutrino can also possess this property, that is, it can coincide with its antiparticle. Such a possibility was first noticed by the Italian scientist Ettore Majorana in 1937. It is still hypothetical and has not been experimentally confirmed. But the contrary option that a neutrino and an antineutrino are dissimilar has not been proven either. To differentiate these alternatives, scientists call a neutrino coinciding with its antiparticle a *Majorana neutrino*. A neutrino having a separate

---

[3]due to an extremely low neutrino mass.

[4]Since right-handed neutrinos almost do not interact with other particles, they can be excellent candidates for objects forming dark matter. Several theoretical models explore this interesting possibility.

antiparticle is called a *Dirac neutrino*. The Standard Model deals only with Dirac neutrinos.

If the neutrino is a Majorana particle, the masses of a left-handed and right-handed neutrino can be different. Alternatively, in the Dirac case, these masses must be the same. Hence, a left-handed and a right-handed Majorana neutrino are, in fact, two distinct particles with different properties. The behavior of left-handed and right-handed neutrinos is dissimilar even in the Standard Model. But if, additionally, neutrinos are Majorana particles, the left- and right-handed neutrinos will also have distinctive *Majorana masses*.

The origin of Majorana mass is not related to the BEH mechanism and must be explained by another theoretical model. Even though this statement is vague and does not clarify too much, it relieves the BEH mechanism from the need to account for the smallness of neutrino mass. Furthermore, it opens the possibility for another theoretical model to fulfill this task. Such a model indeed exists. It is based on two assumptions. It states, first, that neutrinos are Majorana particles, and, second, that the mass of a right-handed neutrino is extremely high. The right-handed neutrino is not observable anyway, so its high mass does not contradict to any experimental result. The model has a funny name: the see-saw mechanism. According to it, the masses of a right-handed and a left-handed neutrino are inversely related: a huge mass of a right-handed neutrino causes a left-handed neutrino to be very light. In some sense, neutrinos play something like the good old children's game. The increase in the mass of one particle effects a decrease in the mass of another.

The see-saw mechanism can explain any small value of the mass of observable left-handed neutrinos. Of course, by doing this, the mechanism requires the mass of undetectable right-handed neutrinos to be incredibly high. The theory does not elucidate the origin of such high mass. Therefore, this invention may look like the replacement of one puzzle by another. Moreover, since a right-handed neutrino can never be exposed, we have no possibility of verifying the see-saw mechanism. So, the predictive power of the theory is reduced. However, there are at least two arguments in its favor.

First, some models of Grand Unification, and, in particular, the SO(10) theory discussed in Section 14.2, do predict the existence of an extremely heavy right-handed neutrino. The expected mass of this particle agrees well with the value wanted by the see-saw mechanism. In other words, this mechanism does not come out of thin air. It is supported by the indirect indications of its validity.

Second, although the see-saw mechanism itself is not testable, the Majorana nature of the neutrino can be experimentally established. Proving that neutrinos are Majorana particles would be a fundamental discovery, which would define the direction of the future development of particle physics. This proof would also strongly support the see-saw mechanism. That is why experiments that aim to ascertain the Majorana hypothesis of the neutrino arrangement deserve separate attention.

Historically, neutrinos were first detected in β decay (see Section 6.1). In this process, a neutron in an atomic nucleus becomes a proton and additionally emits an electron and antineutrino. The corresponding diagram of this process is shown in Fig. 9.4. Some nuclei are subject to double β decay, in which two electrons and two antineutrinos are simultaneously produced. If a neutrino is a Majorana particle, one more process is possible. It is called *neutrinoless double beta decay*, or $0\nu\beta\beta$ decay. As its name suggests, only two electrons without antineutrinos are produced in such decay. Its diagram is shown in Fig. 18.2. In this diagram, the arrow on the neutrino line is intentionally missing, since, for a Majorana neutrino, the particle and antiparticle are the same. $0\nu\beta\beta$ decay is forbidden in the Standard Model because it requires the neutrino to turn into its antiparticle. In other words, an antineutrino must appear in the lower vertex together with the first electron, convert into a neutrino, and after that, produce the second electron in the upper vertex. However, a Majorana neutrino can do this because it coincides with its antiparticle. The Majorana neutrino can be emitted in the lower vertex, where it behaves as an antiparticle, and, after that, can be absorbed in the upper vertex as a particle.

The possibility of $0\nu\beta\beta$ decay was put forward immediately after the seminal work of Majorana. Since then, experimental physicists

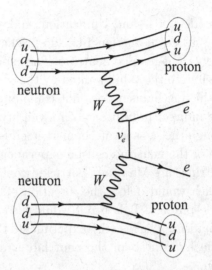

Fig. 18.2   Diagram of neutrinoless double beta decay.

have been trying to detect the process. Contrary to all other decays, neutrinos do not appear in the final state of $0\nu\beta\beta$ decay. Hence, they do not take away a part of the energy. Therefore, the energy of initial and final observable particles (nuclei and electrons) must be exactly equal. This feature allows scientists to distinguish $0\nu\beta\beta$ decay from the usual double β decay, where a considerable part of the energy is carried away by invisible antineutrinos. However, to exploit this feature, physicists need to precisely measure energies of all particles. This task is hard, although scientists have considerably improved the measurement technique during the seventy-odd years of experimental exploration.[5] In some cases, scientists claimed the successful detection of $0\nu\beta\beta$ decay. However, the following studies, more sophisticated and precise, did not confirm the positive result.

All negative results did not weaken interest in this anomalous decay, and its search now continues with commendable perseverance. Moreover, several new experiments boasting superior performance are planned in the future. As always, it is impossible to predict

---

[5]The first search for $0\nu\beta\beta$ decay was attempted in 1948.

whether these efforts will be successful, but this is the usual price to pay for the possibility of unearthing new knowledge about nature.

After developing the see-saw mechanism as an attractive and promising extension of the Standard Model, scientists have found that a right-handed Majorana neutrino can also explain the disappearance of antimatter. The corresponding model of baryogenesis, based on this undetectable particle, is considered in the next chapter.

Chapter 19

Cosmic Drama in Three Acts

In many detective stories, authors employ a simple but very effective plot twist: the least suspicious character suddenly turns out to be the main villain. It may sound odd, but it is quite possible that nature plays the same game with particles, except that, of course, particles cannot be criminals.

We mentioned neutrinos on many occasions in the pages of this book. Each time we stressed that they are "invisible" and "hardly detectable." It may be tough to imagine how neutrinos can be involved in the genesis of matter. They almost do not interact with other particles. So, their influence on what happens in the universe seems practically insignificant. Nevertheless, now we will consider a model of baryogenesis where these invisible particles become the leading stars. As can be supposed, the main emphasis in this model is on leptons, and, because of this, it is called *leptogenesis*.

The first version of leptogenesis was proposed by two Japanese scientists, Masataka Fukugita and Tsutomu Yanagida in 1986. Since then, several modifications of the original scheme were studied by theorists. Today, the uncertainty in the choice of an implementation of leptogenesis still remains. That is why we describe here just the simplest version of the model, which is called *vanilla leptogenesis*, but we also need to mention that many other flavors of leptogenesis exist.

Usually, physical phenomena and art are two non-overlapping areas of human interests. However, some turns in the script of leptogenesis are so unexpected and dramatic that they are reminiscent of a gigantic spectacle played out in a theatre, where the stage is the entire world. That is why we present vanilla leptogenesis as a cosmic drama in three acts. Let us call it *The Genesis of Matter*. The characters of the play in order of appearance are the newborn universe, neutrinos (including a heavy right-handed Majorana neutrino), charged leptons (electron, muon, tauon), quarks, CP violation, thermal equilibrium, sphaleron processes, Higgs boson, electroweak and strong interactions. All of them were discussed in the previous parts of this book. Now, they are united in joint action and participate in a fascinating performance which happened about fourteen billion years ago, and which produced the world where we live.

The central personage, which caused the present imbalance between matter and antimatter, is a very heavy right-handed Majorana neutrino. The selection of this particle for the principal role in the extermination of antimatter may look surprising, because, as we explained before, the Majorana particle and antiparticle are identical objects. However, it is precisely this property of the Majorana neutrino which is the most essential for the whole process of leptogenesis.

The right-handed neutrino does not participate in electroweak and strong interactions. Hence, it is truly invisible and undetectable. Nevertheless, without this particle, nothing would happen. Usually, it is denoted in scientific papers by the letter N, which should signify "heavy neutrino". The same letter was selected by Captain Nemo[1] for his emblem. This coincidence is exceptionally evocative. If any association between particles and personalities of culture is needed, Captain Nemo will be an excellent choice for the N neutrino.

The origin of the large mass of the N neutrino is yet unknown. However, several promising models of Grand Unification naturally predict it. Hopefully, the details of this puzzle will also be collected into a coherent picture sometime in the future. In any case, we should

---

[1]Captain Nemo is a central character in several novels written by the French science fiction writer Jules Verne.

not forget that the high mass of the N neutrino successfully explains an extremely small mass of the observed left-handed neutrinos. But to achieve this, the mass of the N neutrino must be so large that this particle could be created only at the earliest stage of the universe's formation when the temperature was extremely high.

In vanilla leptogenesis, each fermion generation contains an additional N neutrino. This particle can decay and produce a charged lepton, such as an electron or a muon. For example, the possible decay mode can be an electron, an up quark, and a down antiquark. What is important, an N neutrino can also decay to an antilepton because a Majorana neutrino is both a particle and an antiparticle. For example, an N neutrino can decay to a positron, an up antiquark, and a down quark. Hence, the baryon number is conserved in N decays[2] while the lepton number is violated. Decays of the N neutrino should also violate CP-symmetry. So, **Sakharov's second condition must be satisfied at this stage**. The source of CP violation is not specified in the script, but it can be due to differences between direct and reverse transitions of an N neutrino into a charged lepton, that is, something similar to the KM mechanism. No matter what, CP violation in N decays is essential because it must produce an excess of charged antileptons over leptons. Without it, the whole scheme does not work.

Here we need to stress a very peculiar detail. The model of leptogenesis requires that N decays produce an excess of antileptons (that is, antiparticles) over leptons (particles) and not vice versa. In other words, **the first round in the rivalry between matter and antimatter is won by antimatter**. So, the play follows the best traditions of the suspense genre where a superhero loses the first fight with a criminal. However, according to the same traditions, it is not yet the end. As is well-known, one can win a battle but lose a war. In the case of leptogenesis, it means that the initial excess of antiparticles over particles is the main reason for the ultimate dominance of matter in the universe.

---

[2]Each quark has the baryon number $+\frac{1}{3}$, and each antiquark $-\frac{1}{3}$.

Of course, besides decays of N neutrinos, there are also reverse processes, in which charged leptons or antileptons create these particles again. Since both direct and reverse processes are possible, the universe is initially in thermal equilibrium. Hence, any arising excess of antileptons is inevitably washed out. But the universe swiftly expands and, therefore, cools fast. According to theoretical estimates, the cooling rate of the universe is more rapid than the production rate of N neutrinos. Because of this, the universe goes out of thermal equilibrium as was explained in Section 16.1. When this happens, **Sakharov's third condition is also satisfied**, and the produced excess of antileptons is not destroyed anymore.

Eventually, all existing N neutrinos decayed while no new heavy neutrinos could be created. Consequently, these particles disappeared from the stage of the theatre where the cosmic drama was performed many billions of years ago. However, N neutrinos played their part to perfection, and they could proudly leave the scene. N decays violated CP-symmetry, and, simultaneously, they were out of thermal equilibrium. Because of this, an excess of antileptons over leptons was retained in the young universe. In contrast, the number of quarks and antiquarks were still exactly the same. At that point, the first act ended.

In the second act, other actors enter the scene. They are the sphaleron processes. These processes also took place in the first act, but they were in shadow then since they did not influence the main action. But when N neutrinos have disappeared, sphaleron processes take the leading role. The main reason for this is that energy required for sphaleron processes is much smaller in comparison to the production of N neutrinos. Hence, sphaleron processes continue to occur for a long time after the disappearance of N neutrinos. All sphaleron processes have an essential property. They violate both the baryon and lepton numbers. However, **the difference between the baryon and lepton numbers always remains the same**. This property of sphaleron processes was discussed in Section 14.4, and now it starts to play its part in the drama.

The mise en scène for the second act is the following. The universe contains quarks and leptons, which continuously interact, appear and disappear. The only remaining interactions at this stage are electroweak, strong, and Higgs's. The number of quarks and antiquarks at the beginning of the second act is exactly the same. So, the total baryon number of the whole universe is equal to zero. On the contrary, the number of antileptons slightly exceeds the number of leptons, and the total lepton number is negative. Hence, the difference between baryon and lepton numbers is positive. In addition to regular interactions, sphaleron processes are extremely intensive in the second act because the universe is still sufficiently hot, and the energy of particles is reasonably high.

Sphaleron processes fulfill Sakharov's first condition. They turn leptons into quarks and vice versa, and, hence, violate the baryon number. However, they cannot change the difference between the baryon and lepton numbers. For example, they can easily turn two antileptons into nine quarks and one lepton. The corresponding sphaleron process is shown in Fig. 19.1. In the initial state of this process, the baryon number is equal to zero while the lepton number is equal to -2. So, the difference between these figures is $+2$. Each quark has the baryon number $+\frac{1}{3}$. Hence, the baryon number of the final state is $+3$. The lepton number of the final state is equal to $+1$, and the difference between the baryon and lepton numbers is again $+2$. The same arithmetic works everywhere else.

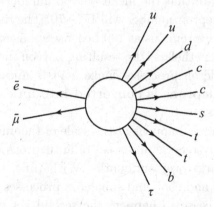

Fig. 19.1 Example of a sphaleron process, which converts two antileptons into nine quarks and one lepton.

Thus, sphaleron processes simultaneously change the baryon and lepton numbers by the same amount. Also, these processes are in thermal equilibrium. As we discussed earlier (see Section 16.3), the high mass of the Higgs boson prevents the departure from thermal equilibrium for the sphaleron processes. This obstacle is a serious problem for electroweak baryogenesis. However, in leptogenesis, the departure from thermal equilibrium takes place in the first act, during decays of heavy N neutrinos. In the second act, during the era of sphaleron processes, it is not needed. Hence, the model of leptogenesis accommodates well the experimental value of the Higgs boson mass.

The thermal equilibrium of the sphaleron processes produces a remarkable effect. The baryon and lepton numbers are modified in each sphaleron process. However, the continuous action of many sphaleron processes for a sufficiently long time settles the average baryon and lepton numbers at well-defined levels. Let us suppose, the physical system initially contains one hundred antileptons and no quarks or antiquarks. That is, the difference between baryon and lepton numbers is equal to 100. The multitude of sphaleron processes, which are in thermal equilibrium, will continuously convert leptons and quarks into each other. But after some time, the average numbers of quarks and leptons in this physical system will attain certain values. After some manipulations with equations, scientists demonstrate that the mean baryon number will be about 30 while the mean lepton number will be -70. The difference between these figures always remains at 100 because sphaleron processes cannot change it. Nonetheless, the resulting baryon number is positive. Hence, the sphaleron processes make a little miracle. **The initial surplus of antileptons is converted into an excess of quarks over antiquarks**.

The same miracle happens at the scale of the universe. Sphaleron processes partially recycle an excess of antileptons over leptons into an excess of quarks over antiquarks. With time, the universe continues to expand and cool, and sphaleron processes gradually diminish. When they cease to happen, the second act comes to an end. Until sphaleron processes completely disappear from the scene, they maintain the average baryon and lepton numbers at certain levels and keep the difference between them equal. So, by the end of the second act, the number of quarks becomes greater than the number of antiquarks.

Here, the plot of the drama is finally revealed, and we must admit that the idea is ingenious. The overall excess of antileptons, which remained after the end of the first act, causes the advantage of quarks over antiquarks at the end of the second act. This gain cannot be removed anymore since none of the remaining interactions can violate the baryon number. Hence, the exceeding number of quarks persists.

What happens in the third act is already well-known. Almost all quarks and antiquarks annihilate. Consequently, antiquarks completely vanish from the universe. However, the small excess of quarks, corresponding to about one additional quark per one billion quark-antiquark pairs, cannot be destroyed. The extra quarks form baryons; baryons unite with electrons and produce atoms; atoms are combined into stardust, and stardust is gathered into stars. The universe, step by step, takes its familiar shape. Finally, it becomes a cozy place where humanity lives and prospers. The curtains come down. The play has a happy ending.

As can be seen, the proposed model of leptogenesis is reasonably simple. It does not require considerable modifications of the well-established Standard Model. Essentially, the main addition of leptogenesis is a heavy right-handed Majorana neutrino. This particle, if it exists, does not influence the Standard Model and its predictions. Moreover, through the see-saw mechanism, a heavy Majorana neutrino relieves the Standard Model of the discomfort caused by the small mass of the observed neutrinos. Hence, the addition of a right-handed Majorana neutrino to the Standard Model seems natural and welcome. As we just realized, this addition also explains the absence of antimatter in the universe.

Thus, the presented script of *The Genesis of Matter* answers the question of missing antimatter. Nevertheless, this successful answer immediately raises another question. Does the dominance of quarks in the universe also imply the excess of antileptons over leptons? This possibility by itself is amazing. We have always been discussing the excess of particles over antiparticles. But the proposed model of leptogenesis claims that the overall number of antiparticles must be higher because of the excess number of antileptons. This claim is contradictory at the first glance. We know perfectly well that the number of charged leptons (that is, electrons) is higher than the number of charged antileptons (positrons) because electrons are included in all atoms. So, where do the antileptons hide?

The answer to this question is quite tricky. We could say that the expected surplus of antileptons is produced by an excess of antineutrinos. It is true that the number of electrons in the universe must be equal to the number of protons because all atoms are neutral. For the same reason, the number of positrons in the universe must be as negligible as the number of antiprotons. In contrast, we know nothing about the relation between the number of antineutrinos and neutrinos. At least, the current knowledge of this relation does not contradict the expected excess of antineutrinos.

However, this is only half of the truth. We should also recall that leptogenesis requires neutrinos to be Majorana particles. Hence, both the neutrino and antineutrino represent two states of the same object. Thus, the final answer of leptogenesis on the imbalance between matter and antimatter is surprisingly beautiful. In this model, the excess of antineutrinos over neutrinos overwhelms the predominance of quarks over antiquarks. Nonetheless, the Majorana nature of neutrinos softens the border between particles and antiparticles and unites the two opposite parts of the microworld into one indivisible whole. So, the fundamental principle of the unity of opposites is nicely confirmed in the relation between matter and antimatter. What a wonderful world our universe is, is it not?

Regardless of all attractive features of leptogenesis, its evidential basis is still slim. However, explorations of this model are being actively pursued. Hence, there is hope that the experimental backing of leptogenesis will solidify in the future. The proof of neutrino oscillations, and, consequently, the evidence that neutrinos are massive, provides an excellent starting point and a strong motivation for this kind of research. Among the possible results, which would support the model of leptogenesis, we can mention the detection of CP violation in lepton transitions and neutrinoless double beta decay. The state of this research changes rapidly, and new results appear each year. Therefore, the probability that leptogenesis will get experimental confirmation in a few years from now is pretty high. Nevertheless, it is also fair to say that, currently, leptogenesis remains an attractive, promising, realistic but still hypothetical model.

Chapter 20

Never-ending Story

This book is coming to an end, and we just need to draw some conclusions. Our main goal was to give an overview of the long path that humankind follows to uncover one of the most intriguing mysteries of nature, the vanishing of antimatter. We attempted to demonstrate how the collective efforts of many thousands of scientists, spanning dozens of years, led to the solution of the problem.

The results of the study of antimatter are really impressive. Of course, scientists have not yet obtained the complete answer. But a significant part of the arduous road has already been covered. Many important pieces of the puzzle have been revealed and understood. Moreover, several credible and realistic theoretical schemes, which can answer the question of missing antimatter, have been developed.

Among these schemes, the most appealing are electroweak baryogenesis and leptogenesis. Each of the models has its attractive features, but, at the same time, both require the presence of new effects, which have not yet been found. In a sense, this fact can be regarded as positive because it means that new discoveries are still ahead of us.

Electroweak baryogenesis is entirely based on the Standard Model. It is a substantial advantage because the Standard Model is a well-tested theory. Many of its predictions, such as the Higgs boson or CP violation, have been experimentally confirmed. Hence, the existence of new phenomena, which are needed for electroweak baryogenesis and which are also predicted by the Standard Model, is very likely. However, scientists already realized that the Standard

Model alone cannot satisfy all requirements of electroweak baryogenesis. Therefore, the addition of new physics is vital for success in this direction.

On the other hand, new phenomena required for implementation of leptogenesis do not change the Standard Model dramatically. They even help the Standard Model to overcome some of the embarrassments related to neutrinos. Without a doubt, such minimal intervention in the configuration of the theory, which has already had a solid reputation for correct description of particle behavior, is a plus for leptogenesis. Besides, this model possesses some intrinsic beauty, which many scientists consider to be an additional condition for validity. If the correctness of the theory of leptogenesis is proven, the end of the search for disappeared antimatter will be comparable to that of classic detective novels. The most invisible and almost undetectable particle, which cannot be suspected to have any significant influence on other objects of the microworld, let alone the destruction of antiparticles, will become the primary object responsible for the genesis of matter.

Whatever model of baryogenesis turns out to be true, it will not be the end of scientific research. As always, resolving one puzzle will just open another hidden door in the maze of nature, and a new mystery, one that is much more cryptic and challenging, will appear in all its enigmatic splendor. There is no end to this journey. We will never uncover all secrets of the universe. Hence, this book is finished, but the story of antimatter never ends. Our travel in the amazing labyrinth of nature goes on.

Index

Printed in the United States
By Bookmasters